CAMBRIDGE LIBRARY COLLECTION

Books of enduring scholarly value

Technology

The focus of this series is engineering, broadly construed. It covers technological innovation from a range of periods and cultures, but centres on the technological achievements of the industrial era in the West, particularly in the nineteenth century, as understood by their contemporaries. Infrastructure is one major focus, covering the building of railways and canals, bridges and tunnels, land drainage, the laying of submarine cables, and the construction of docks and lighthouses. Other key topics include developments in industrial and manufacturing fields such as mining technology, the production of iron and steel, the use of steam power, and chemical processes such as photography and textile dyes.

Memoirs of Richard Lovell Edgeworth, Esq

Richard Lovell Edgeworth (1744–1817) was a noted Irish educationalist, engineer and inventor. This two-volume autobiography, begun in 1808, was published in 1820. Edgeworth had abandoned the project in 1809, having covered the period to 1781, and it was completed after his death by his eldest daughter, a successful novelist. Maria Edgeworth and her father had co-authored educational works, and the experience of helping her father run their estate during her teens had provided material for her novels. Volume 2 of these memoirs was wholly written by her, though it contains excerpts from Richard's correspondence. It recounts how, after his third marriage, the growing family returned to Ireland, and focused first on domestic and educational concerns. Richard became involved in Irish politics and the newly founded Royal Irish Academy but continued to publish essays on scientific and mechanical topics, as well as influential (though controversial) works on education.

Cambridge University Press has long been a pioneer in the reissuing of out-of-print titles from its own backlist, producing digital reprints of books that are still sought after by scholars and students but could not be reprinted economically using traditional technology. The Cambridge Library Collection extends this activity to a wider range of books which are still of importance to researchers and professionals, either for the source material they contain, or as landmarks in the history of their academic discipline.

Drawing from the world-renowned collections in the Cambridge University Library, and guided by the advice of experts in each subject area, Cambridge University Press is using state-of-the-art scanning machines in its own Printing House to capture the content of each book selected for inclusion. The files are processed to give a consistently clear, crisp image, and the books finished to the high quality standard for which the Press is recognised around the world. The latest print-on-demand technology ensures that the books will remain available indefinitely, and that orders for single or multiple copies can quickly be supplied.

The Cambridge Library Collection will bring back to life books of enduring scholarly value (including out-of-copyright works originally issued by other publishers) across a wide range of disciplines in the humanities and social sciences and in science and technology.

Memoirs of Richard Lovell Edgeworth, Esq

Concluded by his Daughter

VOLUME 2

RICHARD LOVELL EDGEWORTH
MARIA EDGEWORTH

CAMBRIDGE UNIVERSITY PRESS

Cambridge, New York, Melbourne, Madrid, Cape Town, Singapore,
São Paolo, Delhi, Dubai, Tokyo, Mexico City

Published in the United States of America by Cambridge University Press, New York

www.cambridge.org
Information on this title: www.cambridge.org/9781108026574

This edition first published 1820
This digitally printed version 2011

ISBN 978-1-108-02657-4 Paperback

Edward. Lord Longford.

Captain. Royal Navy.

Engraved by Henry Meyer from a Picture by Hugh Hamilton.

Published March 30th 1820, by R. Hunter, No. 72. St. Pauls Church Yard London.

MEMOIRS

OF

RICHARD LOVELL EDGEWORTH, ESQ.

BEGUN BY HIMSELF

AND

CONCLUDED BY HIS DAUGHTER,

MARIA EDGEWORTH.

IN TWO VOLUMES.

VOL. II.

LONDON:

PRINTED FOR R. HUNTER,

Successor to Mr. Johnson,

NO. 72, ST. PAUL'S CHURCHYARD,

AND

BALDWIN, CRADOCK, AND JOY,

PATERNOSTER ROW.

1820.

J. M'Creery, Printer,
Took's Court, London.

TO THE READER.

No fact, opinion, or sentiment in the preceding memoirs has been altered by the editor: some verbal corrections, some changes, merely of arrangement, have been made; and a few passages have been omitted, which could be interesting only to the family of the writer.

His manuscript breaks off abruptly; for he was stopped by sudden illness: not that illness which proved fatal; he lived nearly ten years afterwards.

What he wrote of his own life, was written in the years 1808 and 9, at the urgent request of one of his sons, who anxiously desired that he should have continued, and completed his narrative. But after his recovery from the illness by which it had been interrupted, he was called upon to take a part in an active public employment, and his attention was afterwards engrossed by objects, which he thought more useful and important.

Excepting a few passages, he never shewed, even to his own family, what he had written of this account of his life; and when he was urged by them to continue it, he used to say, that "he would leave the rest to be finished by his daughter Maria."

These words struck me always with a sense of dread; yet still I let them pass, with a vain hope that the time was far distant, and that, perhaps, I might, in the chances of human life, never be called upon for the fulfilment of a promise, which had been asked, and given, many years before, when he was in perfect health, and when I was in perfect happiness. In later days, when his health began to decline, he from time to time wrote a few notes of circumstances, which he desired should be recorded: but this was so painful to me, that I could never, even at his request, make any myself, or fix my thoughts upon the subject.—The stroke came upon me the more heavily at last.—After he was no more, I read those solemn and pathetic words, in which he bequeaths the care of his posthumous character "to his beloved daughter," and in which he calls upon me for the performance of a promise and a duty, for which I felt unprepared and unequal.

I resolved,—and it was the only point upon which I could then determine,—that nothing should be written by me hastily. I waited a considerable time, to recover composure of mind. In repeated attempts, I felt how little capable I was of fulfilling the trust reposed in me; but I have persevered. I could not relinquish the hope of doing justice to the memory of my father; of the father who educated me; to whom, under Providence, I owe all of good or happiness I have enjoyed in life. Few, I believe, have ever enjoyed such happiness, or such advantages as I have had in the instructions, society, and unbounded confidence and affection, of such a father and such a friend. He was, in truth, ever since I could think or feel, the first object and motive of my mind.

It may be thought, that with these feelings I am, of all persons, the least fit to be his biographer; and that no time or endeavors can qualify me for the undertaking. The reader will apprehend, that he shall have a panegyric instead of an impartial life and character; he may fear, that he shall be wearied with uninteresting details, or pained by reiterated calls upon sympathy beyond what he can naturally feel: but even if I were not aware, as I am, of the inefficacy and the impolicy of such appeals, I should still shrink from making them: though fully sensible of the generous nature of the British public, yet I could not so far throw myself upon its indulgence. There are feelings, which, conscious of being too strong for public sympathy, will shun observation, and will never solicit compassion.

In continuing these memoirs, I shall endeavor to follow the example, that my father has set me, of simplicity, and of truth. The stamp of truth, which cannot be counterfeited, is seen in what he has written of himself; and will, I trust, be found in whatever I shall add to his narration. The eventful part of his life is ended; nothing more of amusement must be expected from these memoirs. But I may trust to the interest which his narrative has excited, that whoever has followed his history thus far will wish to know more of him. They who have read his account of the circumstances in his early education, which formed his tastes and character, and who may have observed the remarkable influence of one predominant taste, in leading to the events, and to the connexions, both in friendship and marriage, which decided his fate in the beginning of life, will probably desire to follow his course to the end. They will wish to learn what his character became in middle age, and in advanced years; whether he uniformly pursued, or whether he changed, the mode of life he had chosen; and whether ultimately, and upon the whole, it tended to make him happy, or unhappy. Further, they may expect to hear, what claims he may have to public attention as a man of invention, science, or literature: what he attempted or accomplished for the good of his fellow-creatures, either in private or in public life. Those who know any thing of him as a writer on education, may particularly desire to learn what was the result of his opinions, and of his long practice on this important subject. These things, and simply these, I shall state as briefly as possible.

Till now, I have never on any occasion addressed myself to the public alone, and speaking in the first person. This egotism is not only repugnant to my habits, but most painful and melancholy. Formerly I had always a friend and father, who spoke and wrote for me: one who exerted for me all the powers of his strong mind, even to the very last.

Far more than his protecting kindness, I regret, at this moment, the want of his guiding judgment, now when it is most important to me—where *his* fame is at stake.

MARIA EDGEWORTH.

February, 1819.

CHAPTER I.

[1782.

WITH the manuscript of my father's memoirs, I found on the next page to that where his narrative broke off, in his handwriting, the following memorandum.

"In the year 1782, I returned to Ireland, with a firm determination to dedicate the remainder of my life to the improvement of my estate, and to the education of my children; and farther, with the sincere hope of contributing to the melioration of the inhabitants of the country, from which I drew my subsistence."

I accompanied my father to Ireland. Before this time I had not, except during a few months of my childhood, ever been in that country; therefore every thing there was new to me: and though I was then but twelve years old, and though such a length of time has since elapsed, I have retained a clear and strong recollection of our arrival at Edgeworth-Town.

Things and persons are so much improved in Ireland of latter days, that only those, who can remember how they were some thirty or forty years ago, can conceive the variety of domestic grievances, which, in those times, assailed the master of a family, immediately upon his arrival at his Irish home. Wherever he turned his eyes, in or out of his house, damp, dilapidation, waste! appeared. Painting, glazing, roofing, fencing, finishing—all were wanting.

The back yard, and even the front lawn round the windows of the house, were filled with loungers, *followers*, and petitioners; tenants, undertenants, *drivers*, *subagent* and agent, were to have audience; and they all had grievances and secret informations, accusations reciprocating, and quarrels each under each interminable. Alternately as landlord and magistrate, the proprietor of an estate had to listen to perpetual complaints, petty wranglings, and equivocations, in which no human sagacity could discover truth, or award justice. Then came widows and orphans, with tales of distress, and cases of oppression, such as the ear and heart of unhardened humanity could not withstand. And when some of the supplicants were satisfied, fresh expectants appeared with claims of promises, and hopes, beyond what any patience, time, power, or fortune, could satisfy. Such and so great the difficulties appeared to me, by which my father was encompassed on our arrival at home, that I could not conceive how he could get through them, nor could I imagine how these people had ever gone on during his absence. I was with him constantly, and I was amused and interested in seeing how he made his way through these complaints, petitions, and grievances, with decision and despatch; he, all the time, in good humor with the people, and they delighted with him; though he often "rated them roundly," when they stood before him perverse in litigation, helpless in procrastination, detected in cunning, or convicted of falsehood. They saw into his character, almost as soon as he understood theirs. The first remark which I heard whispered aside among the people, with congratulatory looks at each other, was—" His Honor, any way is *good pay*."

It was said of the celebrated King of Prussia, that " he scolded like a trooper, and paid like a prince." Such a man would be liked in Ireland; but there is a higher description of character, which (give them but time to know it) the Irish would infinitely prefer. One who paid, not like a prince, but like a man of sense and humanity. One possessing all the powers of indignant eloquence, yet who used these only when there was just provocation—one excelling in the wit and humor, in which the lower Irish themselves excel; admiring their talents, seeing their faults, loving their generous temper; generous himself, yet not to be easily duped; willing to give, and able to serve, yet neither afraid to refuse nor to punish; humane, but not weak; just, but not to that extreme where justice becomes injustice. Such a man, formed to be loved and respected in Ireland, might hope to be eminently useful in improving the habits and condition of the lower classes of the people.

My father began, where all improvements should begin, at home. He was sensible, that, till his own home was comfortable, he could not pursue his principal objects; he could not set any example of neatness and order, or of propriety and proportion in his mode of living.

His house at Edgeworth-Town, as he mentions in the earlier part of his life, when Mrs. Honora Edgeworth and he lived there, was a tolerably good, old fashioned mansion; but when he now, with seven children returned to it, and considered it with a view to its being the residence of a large family, he felt its many inconveniences.—It had been built in my grandfather's time, in a bad situation, for the sake of presery ing one chimney, that had remained of the former edifice. To this old chimney the new house was sacrificed: to this, and to the fancy, formerly fashionable, of seeing through a number of doors a *suite* of apartments. To gratify this fancy it was made a slice of a house, all front, with rooms opening into each other, through its whole length, without any intervention of passage. All the rooms small and gloomy, with dark wainscots, heavy cornices, little windows, corner chimneys, and a staircase taking up half the house, to the destruction of the upper story. In short, a more hopeless case for an architect, and for a master of a large family, could scarcely occur. It was an immediate temptation to go into grea expense.

At that time, in Ireland, many of the gentry were in remarkable extremes as to their dwelling houses. Some travelled gentlemen erected superb mansions disproportioned to their fortunes, and at last were obliged to sell an estate to pay for a house; or at best they lived in debt, danger, and subterfuge the rest of their days, nominally possessors of a palace, but really in dread of a jail. Others "mistook reverse of wrong for right:" marking these misfortunes, they determined never to build; so lived on in wretched houses out of repair, with locks of doors out of order, the pulleys of windows without sash-line; in short, without what we are accustomed to consider as the common comforts and decencies of life. Others shunned these extremes; but without keeping the safe middle course, they struck out a new half-way mode of going wrong. These would neither build a palace, nor live in a hovel; but they planned the palace, built offices to suit, then turned stable and coach-house into their dwelling house, or provisional residence for the remainder of their days, leaving the rest to fate, and to their sons.

Happily for himself and his family, my father avoided these errors; and he had some merit in so doing, because he had the inventive taste for building, and all the habits of activity and vivacity of temper, which disposed him to be impatient of slow progress towards any object of his wishes. Here, however, he went on by slow degrees, as prudence required, year after year, as his circumstances could afford, without in any one year exceeding his income: he made additions, or alterations, till after exerting some engineering skill, not without frequent predictions from the lookers on, that he would pull the whole house down, he succeeded in making it such as to satisfy his own moderate wishes, and a comfortable residence for a large family.

It was as difficult a task to make any thing of the grounds as of the house: they appeared to have been originally laid out in humble imitation, on a small scale, of the frontispiece to Millar's Gardener's Dictionary, in the original Dutch taste. But the very day after his arrival, he set to work, and continued perseveringly, fencing, draining, levelling, planting, though he knew that all he was doing could not *show* for years. He contented himself with the reflection, that time and industry together might render the whole neat and cheerful. Doomed to a place where nothing sublime or beautiful could be found, he used to comfort himself by considering, that it was the better for his family; he said that, if he had been placed in a charming situation, he might, perhaps, have felt irresistible temptations to expensive improvement.

A celebrated man has declared, that he would ask no better test for judging of the understandings of his fellow-creatures, than the houses they build, and the improvements they make. It is an inadequate, but not altogether an absurd test.

As to society, we had at this time but little, except with Lord Granard's family, at Castle-Forbes, and at Pakenham Hall, the residence of Lord Longford. The connexion and friendship, which had long subsisted between the Pakenham family and ours, my father has mentioned in the early part of his narrative. Had that narrative been continued by him, he would have strongly marked the high regard he felt for Lord Longford, the father of the present Earl. Next to Mr. Day, he was the friend for whom my father had the greatest esteem and affection. Yet Lord Longford and Mr. Day were as different, I might say as opposite, in manners, temper, understanding, habits, and views of life, as any two persons of excellent dispositions and superior abilities, that can be conceived.

Lord Longford, as I have heard him described by my father, was a man of strong practical sense, and decided temper; more for action than debate. Detesting long discussions, he gave his opinion always in few words, and generally with shrewd humour or laconic wit. Little valuing the forms of logical argument, but arriving by some quick, short, and sure mental process at just conclusions, he generally formed right judgments of persons and things; yet was ever unwilling, perhaps unable, to detail in words the reasons, on which his opinions were founded. Enjoying life, without refining or philosophizing, he was good and kind-hearted without parade. In short he had the spirit, principles, and manners of a well-born, well-bred British naval officer. At the time of which I am now writing, just after the American and French war, he was living with his family at home.

His mother, a woman of great wit, and, for her day, of extraordinary knowledge and literature, my father has mentioned with gratitude, as having inspired him, when he was a boy, with an ambition to be something more than a sportsman. Though now considerably advanced in years, her intellectual powers continued in full vigour and animation. Lady Longford, the wife of his friend, was also a woman of quick talents, and of ardent affections. With firm principles, and exalted character, she was fit to be the mother of heroes. A large family was at this time growing up, and educating under her care, at hospitable Pakenham Hall, where various connexions and friends, often drawn happily together, formed a delightful domestic society. But Pakenham Hall was twelve miles distant from us, in the adjoining county of Westmeath. There was a vast Serbonian bog between us; with a bad road, an awkward ferry, and a country so frightful, and so overrun with yellow weeds, that it was aptly called by Mrs. Greville " the yellow dwarf's country."—My father saw but little of the Pakenhams at this period, except in occasional visits.

Castle-Forbes, the residence of the Earl of Granard, was more within our reach than Pakenham Hall; still it was eight or nine Irish miles distant. There the society was various, and very agreeable, especially when Lady Granard's mother, the late Lady Moira, was in the country. Lady Moira's taste for literature, general knowledge, and great conversational talents, drew round her cultivated and distinguished persons; but it was her noble, high spirited character, which struck my father still more than her acquirements or her abilities. Eager to give his family opportunities of enjoying the advantages of such society, he was gratified by the manner, in which she first encouraged and distinguished his daughter, and grateful for the friendship, with which Lady Moira honored her ever afterwards. Such kindness he always felt more strongly, than any that could be shewn to himself.

In our more immediate neighbourhood, we at this time commenced an acquaintance with a friendly and cultivated family of the name of Brooke. The father, an old, well-informed clergyman, was nearly related to the Mr. Brooke, who wrote the celebrated novel of "the Fool of Quality," and the tragedy of "Gustavus Vasa." He possessed a considerable share of his relation's original genius, enthusiasm, and simplicity of character. With much classical learning, he had an admiration for Homer, which he expressed often with a vehemence, that appeared extravagant in the opinion of his common auditors, but in which my father most cordially sympathised. Mr. Brooke's daughter was married to Mr. Eyles Irwin, the well known traveller; so that by another author this family were connected with the literary world.

Considering the state of society in Ireland, at the time of which I am now writing, my father may be esteemed fortunate, in finding in a remote place such acquaintance. In general, formal large dinners and *long sittings* were the order of the day and night. The fashion for literature had not commenced, and people rather shunned than courted the acquaintance of those, who were suspected to have literary taste or talents. But even at that time my father foresaw, and foretold, the happy change, which increasing intercourse with other countries, improving education, and the consequent progress of the diffusion of knowledge, would in a few years produce in Irish society.

CHAPTER II.

THE advantage of the residence of proprietors upon their estates in Ireland can be fully understood only from the example or evidence of those, who have had actual experience on the subject. In this point of view, accounts of the affairs of private Irish gentlemen may be useful to the public. The value of the experience must be measured by the accuracy of the evidence, and by the extent of the opportunities for acquiring information. It is not essential, that the observations should have been made on a large scale as to property, provided that their accuracy has been verified by time.

From the period of my father's settling in Ireland, in 1782, he resided in that country, and on his own estate, nearly five and thirty years. I never was absent from him, during the whole of that period, excepting at the utmost three or four months. —Therefore, as to length of time, my evidence is good; and as to opportunities for observation, none could be greater than those I enjoyed. Some men live with their family, without letting them know their affairs; and however great may be their affection and esteem for their wives and children, think that they have nothing to do with business.—This was not my father's way of thinking.—On the contrary, not only his wife, but his children knew all his affairs. Whatever business he had to do was done in the midst of his family, usually in the common sitting-room: so that we were intimately acquainted, not only with his general principles of conduct, but with the most minute details of their every-day application. I further enjoyed some peculiar advantages:—he kindly wished to give me habits of business; and for this purpose, allowed me during many years to assist him in copying his letters of business, and in receiving his rents.—Consequently, I can now state, not merely my hearsay opinion and belief, but my evidence, with regard to his conduct as a landlord.

As soon as he returned to Ireland, he began to receive his rents without the intervention of agent or sub-agent. On most Irish estates there is, or there was, a sort of personage commonly called a *driver;* a person who drives and impounds cattle for rent and arrears. Such persons being often ill chosen, and of the lowest habits as well as of the lowest order, misuse their authority; and, frequently unfaithful to the landlord, as well as harassing to the tenant, sell the interest of their employer for glasses of whiskey; and finish by running away with money, which they have received *on account,* or by extortion from tenants. These *drivers* are, alas! from time to time too necessary in collecting Irish rents.—My father, however, took from his, all discretionary power, and rendered this petty tyrant's authority as brief as possible. He forbad him to call upon any tenants without express orders, or to threaten seizure of cattle, or upon any pretence to receive money from them *on account.*

He desired, that none of his tenants would pay rent to any person but to himself, or at his own house; thus taking away subordinate interference, he became individually acquainted with his tenantry—saw, heard, talked to them, and obtained full knowledge of their circumstances and characters. From his own observation, and from inquiry from the most intelligent farmers, he soon made himself master of the value of the different land on his estate.

Considerable farms fell out of lease; and he was then able to fulfil his wish of doing what was both just and kind. In every case where the tenant had improved the land, or even where he had been industrious, though unsuccessful, his claim to preference over every new proposer, his *tenant's right*, as it is called, was admitted.—But the mere plea of "*I have lived under your Honor, or your Honor's father or grandfather,*" or "*I have been on your Honor's estate so many years*"—he disregarded. Farms, originally sufficient for the comfortable maintenance of a man, his wife, and family, had, in many cases, been subdivided from generation to generation; the father giving a bit of the land to each son to settle him.—The hope was, that these settlements would be confirmed by the landlord at the expiration of the lease; but this he would not do. The maintenance was hardly sufficient to keep them one step above beggary; and insufficient even for this, when the number of their children increased. It was an absolute impossibility, that the land should ever be improved, if let in these miserable *lots*. Nor was it necessary that each son should hold land, or advantageous that each should live on *his* " *little potato garden*," without further exertion of mind or body.

Parts of Ireland were at this time, and I believe still are, in the same state in this respect as some parts of Scotland, described by Lord Selkirk in his excellent Essay on Emigration; and a case somewhat similar is represented in a late well written tract on the improvements in the Highlands on the Sutherland estate, made by the Marquis and Marchioness of Stafford. Except where there is, as in their instance, an almost princely extent of territory, and vast resources of wealth and power, joined to benevolence, judgment, and activity in the management of resources, little can be done by any private individual, to alleviate the misery necessary in the passage from one state of civilization to another. But whatever any one proprietor *can* do, he ought to attempt; because, however small the actual benefit, his example may influence others; and the junction of many individuals may be of beneficial consequence, either in resisting prejudice, or in giving effect to judicious benevolence.

Not being in want of ready money, my father was not obliged to let his land to the highest bidder, and at what is termed the highest penny. He could afford to have good tenants; and in selecting these, he was not merely decided by their being *substantial* persons, but by their having good characters. He was not influenced by party prejudices, or electioneering interest, to receive or reject those of any persuasion, merely on account of their religious tenets. From the reasonable rate at which he let his land, he found sufficient competition among tenants, to enable him to make any conditions that he thought proper in his leases. He never made any oppressive claims *of duty fowl—of duty work, of man or beast.*—In the old leases made in his father's time, such had been inserted; but he never claimed or would he accept of them, though such were at that time common. He was, I believe, one of the first to abolish them. He refused also to let leases to tenants in partnership, or tenants in common, a mode of tenure which subjects the industrious and skilful to suffer for the faults of their idle or vicious copartners.

Observing that much harm is done by those country gentlemen, who, ignorant of the principles of political economy, with arbitrary or superfluous interference endeavour by bounties or restrictions, to prevent or forward the natural course of things, my father abstained from all such interference. The more he read and thought upon the subject, the more he was convinced of its folly and danger. For example, he never attempted to force markets or manufactures, or to raise or lower wages of labor. He never even bound his tenants to have their corn ground at any particular mill, —a common restriction in an Irish lease.

He omitted a variety of old feudal remains of fines and penalties: but there was one clause, which he continued in every lease, with a penalty annexed to it, called *an alienation fine.*—A fine of so much an acre upon the tenant's reletting any part of the devised land. This clause he thought necessary for several reasons, both for the interest of landlord and tenant. To protect the landlord from the danger of having his property pass from the hands of good and agreeable tenants, to those who might be litigious, or insufficient; to prevent a set of *middlemen* from taking land at a reasonable rent, and reletting it immediately to poor tenants at the highest price possible to be obtained from their necessities. The evils and misery resulting from that system are now sufficiently understood. But when they were not as clearly seen as they now are, my father exerted himself to withstand and oppose them. He never failed to insist upon the payment of the alienation fine, in every one of the few instances, in which any of his tenants braved the clause. The fine was generally calculated so nearly to the value of the utmost additional rent, that could be obtained by reletting the land, that the tenant found it not for his interest to disobey.

The oppression and distress, to which the wretched undertenants were often in other cases subject, will scarcely be believed. It happened, not unfrequently, that the first tenant, the *middleman*, being either fraudulent or extravagant, unable or unwilling to pay, the landlord had no resource, but, in the technical phrase, *to go to the land.*—That is, to send the driver to seize whatever cattle, or produce, could be found on the farm, and to sell these for rent. Now the middleman either having no stock, or having taken care in time to remove it, the loss fell upon the poor undertenants, who often had paid him their rent, yet were nevertheless obliged to pay that rent over again to the head landlord, or to be ruined by the sale of their cattle and goods at inadequate prices. Instances of this horrible injustice were frequent. Nor was it in the head landlord's power, at that stage of the business, to do otherwise.—What could he do?—He saw before him, perhaps for the first time in his life, a set of poor wretches, undertenants, who had come upon his land without his consent or knowledge. His heart might be touched by their misery; but his interest, his own necessities, were still to be considered. He had no other means of obtaining his rent, but by *coming upon them.* An act of parliament for the protection of Irish under-tenants, enabling them, by an easy process, to recover from the middle landlord, whatever, on being driven, they might have been forced unjustly to pay to the head landlord, passed in 1817,—the last year of my father's life.

As to the time or term of his leases, desiring that they should be such as to afford equal security and equal share of advantage to landlord and tenant, my father would have preferred Lord Kaimes's form of covenant to all others. In which, after a certain term of years, (seven, I believe) the lease is renewable or void, at the option of the tenant or landlord, on the valuation of the tenant's improvements, and on a proportional rise of rent being offered. But in adopting, or endeavoring to introduce any unusual form of lease, difficulties occur, arising from the local circumstances and previous habits of the people. There was not in Ireland sufficient confidence between landlord and tenant, nor sufficient habit of——what shall I say?—I must not say *honesty*,—but accuracy—punctuality in their dealings, to admit of such a lease. There was a continual struggle between landlord and tenant, upon the question of long and short leases. Much eloquence was exerted by the tenant, to convince the landlord's understanding, that it was better to give a long lease; because, as he said, no tenant could otherwise afford to improve. The offer of immediate high rent, or of *fines to be paid down* directly, tempted the landlord's extravagance, or supplied his present necessities, at the expense of his future interests; and though aware, that the value of improvable land must rise, or that he was letting it under its actual value; yet if the landlord was not resident on his estate, and if he merely wanted to get his rents without trouble, he was easily tempted to this imprudence. Many have let for ninety-nine years; and others, according to a form common in Ireland, for three lives, renewable for ever, paying a small fine on the insertion of a new life, at the failure of each. These leases, in course of years, have been found extremely disadvantageous to the landlord, the property having risen so much in value, that the original rent was absurdly disproportioned. They have frequently tried to take advantage of certain clauses usual in these leases, by which the lease is void, if the renewals have not been executed or claimed by the tenant, or if the fines have not been paid within a certain time. This opens a field for perpetual litigation or ill-will, consequent on difference of interest between landlord and tenant. Several such leases came down to my father from his father and grandfather, especially of houses in Edgeworth-Town —It was thought, that such were necessary to induce persons to build.—He never made any such leases himself, believing it to be one of the most bungling methods of giving away property. At the moment, it may appear to the improvident individual much the same, whether he give a lease renewable for ever, or for ninety-nine years, as he is not likely to see the end of this term; but to his family it is of very different importance. My father, in the course of his life, saw the end of two leases for ninety-nine years, granted by his predecessors, and enjoyed in consequence considerable rise of property. In these, and all cases where long leases had been granted, he did not find, that the land had been improved by the tenants, or that they felt any gratitude, for what had been originally desired and granted as a favor. On the contrary, long possession had made the occupier almost forget, that he was a tenant; and consider his being forced to surrender the land at the expiration of the lease as a great hardship. The longest term my father ever gave was thirty-one years, with one or sometimes two lives. He usually gave one life, reserving to himself the option of adding another, the son, perhaps, of the tenant, if he saw that the tenant deserved it by his conduct. This sort of power to encourage and reward, in the hands of the landlord, is advantageous in Ireland. It acts as a motive for exertion; it keeps up the connexion and dependence, which there ought to be between the different ranks, without creating any servile habits, or leaving the improving tenant insecure as to the fair reward of his industry.

Those who have tried to do good as legislators, with the most extensive views—or as private individuals, in the management of their own concerns on the smallest scale—must know, that in endeavoring to abolish any abuse of long standing, or even to restore rational liberty to those who have been in any instance long oppressed, there occurs often some dangerous reaction,—some injustice to others, which had not been foreseen. This cannot always be provided against in cases of sudden abolition. So that it is both prudent and benevolent, to take not that, which, abstractedly viewed, is the best possible course, but that which is the best the circumstances will altogether allow.

When the oppressive duty-work in Ireland was no longer claimed, and no longer inserted in Irish leases, there arose a difficulty to gentlemen in getting laborers at certain times of the year, when all are anxious to work for themselves: for instance, at the seasons for cutting turf, setting potatoes, and getting home the harvest.

To provide against this difficulty, landlords adopted a system of taking duty-work, in fact, in a new form. They had *cottiers* (cottagers), day-laborers established in cottages, on their estate, usually near their own residence: many of these cabins were the poorest habitations that can be imagined; and these were given *rent free,* that is, the rent was to be *worked* out on whatever days, or on whatever occasions, it was called for. The grazing for *the* cow, the patch of land for flax, and the ridge or ridges of potato land were also to be paid for in days' labor in the same manner. The uncertainty of this tenure *at will,* that is, at the pleasure of the landlord, with the rent in labor and time, variable also at his pleasure or convenience, became rather more injurious to the tenant, than the former fixed mode of sacrificing so many days' duty-work, even at the most hazardous seasons of the year.

My father wished to have entirely avoided this cottager-system; and he at first thought, by giving increased and sufficient wages, he could secure laborers whenever he might want them, without keeping men in a state of dependence too nearly approaching to slavery. But *total abolition* was not at that time expedient, or practicable. When an individual raised the usual price of labor, it produced but a partial effect, and did not secure the intended object. The laborers flocked to him at the season when they were not wanted; and just at the time when they were wanted, went to any one, who at that moment gave higher wages. There was not any sufficient hold of sentiment or gratitude, that could stand against small pecuniary temptations; nor could the demonstrated view of their own permanent interest avail. My father was obliged to adopt a middle course. To his laborers he gave comfortable cottages at a low rent, to be held at will from year to year; but he paid them wages exactly the same as what they could obtain elsewhere. Thus they were partly free, and partly bound. They worked as free laborers; but they were obliged to work, that they might pay their rent. And their houses being better, and other advantages greater, than they could obtain elsewhere, they had a motive for industry and punctuality; thus their services and their attachment were properly secured. Under the words *other advantages* are included potato land, and what are termed *con acres* (corn acres), flax ground, &c. These put means of just reward and punishment into the landlord's hands at small expense, but in the hands of a venal steward or under-agent, not properly attended to, they become sources of the worst extortion and cruelty.

My father's indulgence as to the time he allowed his tenantry for the payment of their rent was unusually great. He left always a year's rent in their hands: this was half a year more time than almost any other gentleman in our part of the country allowed. Some landlords made their tenants pay up to the day; or, in their mode of expressing themselves, made them English tenants. How far his indulgence was advantageous or hurtful to his tenantry, and to his own interests, I shall not yet stop to inquire. Many years' experience must be taken, to decide such a point; and we are now only stating the facts and the principles, upon which he acted. The rents were paid half yearly: he was always very exact in requiring, that the tenants should not, in their payments, pass beyond the half-yearly days—the 25th of March and 29th of September. In this point they knew his strictness so well, that they seldom ventured to go into arrear, and never did so with impunity.

To guard against his own good nature and generosity is a difficulty, which every generous or good-natured man must soon feel in settling in Ireland; my father felt it early and late. If the people had found or suspected him to be weak, or, as they call it, *easy*, there would have been an end of all hope of really doing them good. They would have cheated, loved, and despised a mere *easy* landlord; and his property would have gone to ruin, without either permanently bettering their interests or their morals. He, therefore, took especial care, that they should be convinced of his strictness in punishing, as well as of his desire to reward. Frequently he spoke with more severity than he felt. This they had penetration enough to see through immediately, and they comprehended his motive. —Often I have heard his old steward say, with an expressive smile and shrug, "if "my master were but half as bad as he says, "it would be bad enough for us; but he's "only afraid of himself being too good,— "and he's right enough there."—Examples of firmness were necessary. These were given publicly, and chosen well; that is to say, they were chosen from cases where the sympathy of the by-standers was clearly against the culprit; for example, where his ingratitude and treachery were notorious.

Where the offender was tenant, and the punisher landlord, it rarely happened, even if the law reached the delinquent, that public opinion sided with public justice. In Ireland, it has been, time immemorial, common with tenants, who have had advantageous bargains, and who have no hopes of getting their leases renewed, to waste the ground as much as possible; to *break it up* toward the end of the term; or to *over-hold,* that is, to keep possession of the land, refusing to deliver it up. This obliged the landlord to a tiresome law process, and subjected him to lie out of his rent for at least a twelvemonth.

A tenant, who held a farm of considerable value, when his lease was out, besought my father to permit him to remain on the farm for another year, pleading, that he had no other place, to which he could at that season, it being winter, remove his large family. The permission was granted; but at the end of the year, taking advantage of this favor, he refused to give up the land. Proceedings at law were immediately commenced against him; and it was in this case, that the first trial in Ireland was brought, on an act for recovering double rent from a tenant for holding forcible possession, after notice to quit.

This vexatious and unjust practice of tenants against landlords had been too common, and had too long been favored by the party spirit of juries; who, being chiefly composed of tenants, had made it a common cause, and a principle, if it could in any way be avoided, never to give a verdict, as they said, against themselves. But, in this case, the indulgent character of the landlord, combined with the ability and eloquence of his advocate*, succeeded in moving the jury,—a verdict was obtained for the landlord. The double rent was paid; and the fraudulent tenant was obliged to quit the country *unpitied.* Real good was done by this example: it was seen, that he could on proper occasions resist his natural disposition to forgiveness and compassion; his character for firmness was thus established; and consequently he was respected as well as beloved, and the more loved by the common people for being respected.

There were some other points necessary to be established on coming to settle in Ireland. Gentlemen proprietors, often from the spirit of party and of pride, or from electioneering motives, or pecuniary interest, made it a practice not only to skreen, but, as they call it, *to protect* their tenants, when they got into any difficulties by disobeying the laws. Smuggling and illicit distilling seemed to be privileged cases, where, the justice and expe-

* Mr. Fox, afterwards Judge Fox.

diency of the spirit of the law being doubtful, escaping from the letter of it appeared but a trial of ingenuity or *luck*. In cases that admitted of less doubt, in the frequent breach of the peace from quarrels at fairs, rescuing of cattle driven for rent, or in other more serious outrages, tenants still looked to their landlord for protection; and hoped, even to the last, that his Honor's or his Lordship's interest would get the fine taken off, the term of imprisonment shortened, or the condemned criminal snatched from execution. In a country, and in a state of society, where habits and feelings such as these prevailed, and had been long established; and where these ideas of protection and favor constituted almost an essential part of the idea of a good landlord; it was difficult for any individual, to act directly contrary to the prepossessions, habits, and interests of the common people, and yet to preserve their attachment: but this my father effected. He never would, on any occasion, or for the persons he was known to like best, interfere to *protect*, as it is called, that is, to skreen, or to obtain pardon for any one of his tenants or dependants, if they had really infringed the laws, or had deserved punishment.

He went counter to their prejudices, hopes, and interests, in other particulars. He never would exert himself, to obtain certain little *preferences*, which those gentlemen, who have influence on county grand juries, and who can satisfy their consciences that *jobbing* is right and necessary, can easily obtain. Without descending into particulars too minute or invidious, it may just be suggested, as it has lately appeared in evidence before the House of Commons, that much public money is wasted in road-making, and in other country works in Ireland. It has been said in excuse, that this money affords tenants the means of paying their rents. My father's tenants were never favored in this manner, and yet they paid their rents. He set an example of being scrupulous to the most exact degree as a grand juror, both as to the money required for roads or for any public works, and as to the manner in which it was laid out. At first this diminished both his popularity and his influence, because it decreased his power of giving certain customary rewards. But it was soon overbalanced by other circumstances, and by his having other rewards at his disposal.

It has been observed, that, by the land which he kept in his own hands, he had continual means of serving and obliging the deserving poor in his neighbourhood; he employed in his buildings and improvements a great number of laborers and workmen of various sorts—these found him a punctual, prompt, and liberal paymaster, a judge of their different merits, and a person able and willing to forward them in the world.

To his character as a good landlord was soon added, that he was a *real gentleman.* This phrase, pronounced with well known emphasis, comprises a great deal in the opinion of the lower Irish. They seem to have an instinct for the *real gentleman,* whom they distinguish, if not at first sight, infallibly at first hearing, from every pretender to the character. They observe, that the real gentleman bears himself most kindly, is always the most civil in speech, and ever seems the most *tender of the poor.*

This good seeming, they found, was something more than *seeming* in my father. They soon began to rely upon his justice as a magistrate. This is a point, where, their interest being nearly concerned, they are wonderfully quick and clearsighted; they soon discovered, that Mr. Edgeworth leaned neither to Protestant nor Catholic, to Presbyterian nor Methodist; that he was not the *favorer* nor partial protector of his own or any other man's followers. They found, that the law of the land was not in his hands an instrument of oppression, or pretence for partiality. They discerned, that he did even justice; neither inclining to the people, for the sake of popularity; nor to the aristocracy, for the sake of power. This was a thing so unusual, that they could at first hardly believe, that it was really what they saw. Electioneering motives, and the secret action of personal friendship and aversion, were of course habitually suspected. But, the plain facts forced their way, "Go before Mr. Edgeworth, and you will surely get justice," was soon the saying of the neighbourhood. Besides relying on his justice, they felt with all the warmth of their warm hearts his eagerness, to exert himself in the cause of the injured or oppressed. This touches the Irish more than all the rest, they really are more attached by what touches their hearts, than by what concerns only their interests.

Altogether he was fit to live in Ireland, and to accomplish his own wish of meliorating the condition of the people. In this endeavor he found it necessary to go very gradually to work, by exciting the wish for improvement in the first place, and by taking only the smallest step at a time, beyond what they had been used to, as to cleanliness, neatness, and order. He suggested rather than commanded, and assisted those only who had shewn by some effort, that they were willing to assist themselves. The belief among the lower Irish, that the landlord must have favorites, and must do all the good he does from *favor* and *affection,* was not to be suddenly changed by any strictness of impartiality, or demonstration of justice. It was better to make use of their prejudices, and he did so.

For instance, with regard to drinking.

Instead of trying to convince their understanding of its injurious consequences, he inveighed against drunkenness with eloquence suited to their comprehension; declaring, that his *favor* should never be shewn to any man, whom he should see drunk. Thus impressing upon them the belief, that "the *master* had an antipathy to drinking;" and that, to use their own phraseology, "if the *master* should once *catch* a man "drunk, he would never forget or forgive it "to him or his;" operated more powerfully, than any moral or rational argument, that could have been used. Bad habits can scarcely be broken, even with the most rational people, by the mere power of conviction. It was necessary, first to produce a striking effect; afterwards, by degrees, as a new generation, better educated, and more enlightened, should arise, he knew, that he might make himself better understood, and could gradually substitute reasoning for authority.

In his attempts to make any improvement in managing their domestic business, or their moral economy, he used example more than precept, exhortation, or authority.

Soon after his return to Ireland, he set about improving a considerable tract of land, reletting it at an advanced rent, which gave the actual monied measure of his skill and success. This raised his credit with the farmers, and with the lower class of people: his advice upon such subjects was consequently really desired; not merely admitted or assented to with the hypocritical, "Oh, true your Honor;" or, "Oh, sure your Honor must know best."

I have endeavoured to state the general principles upon which he acted, rather than to give particular instances of the good he did or attempted. I know, that it was his opinion, that more good can be effected, by adhering uniformly to right principles, than by any particular acts of what may be called generosity or charity. Besides, it would be impossible for me, without ostentation or indelicacy, to give any of the proofs I might record of my father's liberality. Long after they were forgotten by himself, they were remembered by the warm hearted people among whom he lived; and now he is no more, they will be recollected by some who read this, with what he would prefer to all thanks, or to any praise—heart-felt gratitude and affection.

CHAPTER III.

[1782.

Before we go on with his domestic life, it is necessary to give an account of some public political transactions, in which my father was called upon to take a part, immediately upon his arrival in Ireland. He has left a few pages of memorandums on this subject, but scarcely sufficient to be intelligible to those, who are not acquainted with the Irish politics of that day. It is a subject foreign to my habits, and on which I should not touch for any slight consideration; but I feel it my duty to state whatever is necessary for the thorough understanding the conduct and opinions of him, whose life I am writing.

In 1782, when he returned to Ireland, the Irish volunteers were in force. The history of their rise and progress, though it must be familiar to every Irishman, may not be quite so well known to others. They first appeared in 1778, at a disastrous period of the American war, when England, obliged to employ her armies elsewhere, had not troops sufficient for the defence of Ireland, which was then threatened with invasion, and exposed to domestic insurrection. Her principal nobility and gentry embodied themselves volunteers for the defence of the country, and the first corps was under the command of the Duke of Leinster. In time they increased to a body of about fifty thousand men, under the command of captains, colonels, and a general of their own chusing. The Earl of Charlemont was their general. Their existence and their proceedings were neither owned nor disavowed by the legislature, till one of the Lord Lieutenants of Ireland, at a moment of danger, when the north was threatened with invasion, and no other troops were to be had, applied to Lord Charlemont for the assistance of the volunteers. Continuing for several years under perfect order and good discipline, they rendered their country respectable in the eyes of England, as well as of foreigners, and more than suggested the possibility of her independence. Irish patriots felt the advantages, that might be made of their position, and they used their power with prudence and energy. The friends of Ireland, both in the Irish and English parliament, obtained for her great benefits; some for her commerce, more for her constitution and liberty. Between the years 1778 and 1783, many laws obnoxious to Ireland were repealed; in particular, one penal statute against the catholics, which we can now perhaps scarcely believe to have existed in our own times. It was a statute, of which the penalties had sometimes lain dormant, but had, at others, been called into action for the purposes of private treachery, or public persecution.

A statute of King William the Third, entitled, an act to prevent the growth of popery, ordained no less than forfeiture of inheritance against those catholics, who had been educated abroad; at the pleasure of any informer, it confiscated their estates to the next protestant heir: that statute further deprived papists of the power of obtaining any legal property by purchase; and, simply for officiating in the service of his religion, any Catholic priest was liable to be imprisoned for life. Some of these penalties had fallen into disuse; but, as Mr. Dunning stated to the English House of Commons, "many respectable Catholics still lived in fear of them, and some actually paid contributions to persons, who, on the strength of this act, threatened them with prosecutions." Lord Shelburne stated, in the House of Lords, "that even the most odious part of this statute had been recently acted upon in the case of one Molony, an Irish priest, who had been informed against, apprehended, convicted, and committed to prison, by means of the lowest and most despicable of mankind, a common informing constable. The privy council used efforts in behalf of the prisoner; but, in consequence of the written law, the king himself could not give a pardon, and the prisoner must have died in jail, if Lord Shelburne and his colleagues had not released him at their own risk."

In moving for the repeal of this law, Sir George Saville declared, "that one of his principal views was to vindicate the honor, and assert the purity of the Christian religion." The motion for its repeal passed the English House of Commons without a single negative, and in the House of Lords it was opposed but by the solitary voice of one bishop. Thus, Lord Nugent, Sir George Saville, Mr. Burke, Mr. Dunning, in the English House of Commons, and Lord Rockingham and Lord Shelburne, in the House of Lords, by the combined force of patriotism, integrity, law, reason, and political ability, obtained this first step towards the improvement of the condition of the Catholics in Ireland. The Irish patriots, encouraged by success, and by their powerful position, turned to other objects, raised their tone, and called loudly for general emancipation from the legislative and commercial restrictions of England. In 1781, the celebrated convention of delegates, from one hundred and forty-three coups of volunteers, assembled at Dungannon. Their resolutions were adopted also by the volunteers of the South; and at length

Mr. Grattan moved an address to the throne, asserting the legislative independence of Ireland. The address passed; the repeal of a certain act, empowering England to legislate for Ireland, followed; and the legislative independence of this country was acknowledged. This kingdom was then in a transport of joy, and the patriots and volunteers were at the height of their popularity.

We landed in Dublin at the moment when the nation voted a parliamentary grant of fifty thousand pounds, as a testimony of their sense of the services of their great and successful orator and patriot Mr. Grattan. My father was in the gallery of the Irish House of Commons, when that vote passed, and he saw and sympathised with the public enthusiasm. In the midst of this enthusiasm, and while the volunteer patriots, in the general exaltation of their spirits, anticipated a variety of future undefined advantages to Ireland, his hopes fixed on that, which in his view was the most important object, and without which all others, as he thought he foresaw, would be of no avail. A few days after he came to Dublin, he wrote and published the following address to the associated volunteer corps of the county of Longford :—

" Gentlemen,

" I congratulate you upon the success of your manly and prudent conduct: from a state of dependence and danger you are become secure and free;—no foreign enemy can alarm you, no jealous rival refuse you the rights of Freemen.

" But, at this moment of universal joy and exultation, let me recall to your remembrance, that without an effectual Reformation of the Irish House of Commons, no solid advantage can be obtained from our present success. A venal Parliament may, by degrees, yield every thing but the name of Freedom; and the slow but certain influence of corruption may, in a few years, reduce you to the same subjection, which you have so nobly shaken off by the well-timed exertions of national courage, and unexampled prudence.

" Every volunteer at this instant feels, and exults in his own consequence: did he feel this ardor for liberty, this independence of rank and wealth, before he became a volunteer? Does he expect, that, when the corps he belongs to disbands, or forms part of a militia, he will hold the same share of the national power, which he does at present?—No, the instant he lays down his arms, and that instant must arrive, his consequence and all his rights as a man and citizen fall into the hands of a few, who influence the boroughs, which return the majority of the Parliament. Here, my friends, is the object, upon which the real and permanent happiness of your country depends: increase the number of representatives for your counties, shorten the period of their delegation, balance the parliamentary weight of your a broughs, and every avenue of your freedom will be guarded. From a parliament dependent only on its constituents every measure of public utility and internal prosperity will necessarily arise, and every freeman in Ireland will possess his just share in the legislation of his country. It is not for me to point out the means or measure of this important reformation; the wisdom of all the associated corps of Ireland must be combined to determine them. But, as a friend to my country, I think it my duty, at this crisis of your fortune, to urge the only measure, which I believe to be *effectual* for your future happiness and dignity. The liberality of the volunteer resolutions respecting religious toleration, and the late act in favor of the Roman Catholics of this country, tend to a union of interests, that promises the most favorable consequences. What can promote this approach of sentiments more strongly than increasing the importance of the rights of freeholders —Their importance will not merely be *doubled,* by doubling the number of county members—it will increase in a proportion greater than I will venture to determine. No little interest between landlord and tenant will then subjugate the latter to a blind dependence on the owner of his farm: a yeomanry will, by degrees, arise, which will diffuse liberty and industry through every class of the inhabitants of Ireland.

" The aristocratic, by far the most powerful part of this kingdom, would at any other time smile at an attempt upon its influence. Trust me, my friends, that aristocracy must restore to you your rights, if you demand them; and the time will come, when even the families, which now possess the most extensive interest in boroughs, will bless the hour, that turns their ambition from a sordid scramble for titles and places, to the solid happiness of serving an industrious and prosperous people.

"For myself, I have no interest to serve, no private pique to gratify; nor have I any view in offering you my advice, but what every Irishman would be proud to avow—the liberty and glory of my country.

"I hope the importance of this subject will induce you, to assemble and deliberate upon its expediency, though it is not offered to your consideration by a person of more consequence than

"Your obedient and faithful servant,

"RICHARD LOVELL EDGEWORTH."

The first copy of this address was, as I remember, lying on my father's table, in Dublin, when Mr. William Foster, (afterwards bishop of Clogher,) his old schoolfellow and friend, came to see him. After looking over it, he smiled, with his look of good humoured raillery, took up a pen, and wrote the word *free* before the signature, in the manner in which members of Parliament frank letters; inferring from the popular topics and tone of the address, that the writer was preparing to become a candidate for the county of Longford. But his friend was mistaken, neither popularity, nor a seat in parliament, was my father's object; he meant only that, which persons, who have seen much of the political world, can scarcely believe, that a man past thirty, and a man of talents, can intend by a political address, *simply* the good of his country. The expectation, that he would declare himself a candidate, was raised almost to a certainty, when he appeared at a county meeting, called for the purpose of naming candidates.

At this meeting he proposed, and with some difficulty carried resolutions, and a petition for parliamentary reform. These were the first resolutions, and this was the first petition on that subject in Ireland. Other counties eagerly followed this example; and in the course of a few months there were county meetings, and meetings of volunteer corps, all over Ireland: delegates from the volunteers were appointed to provincial assemblies, whence a fresh selection was to be made for a grand national convention in Dublin. My father, of course, was a delegate to the Leinster provincial meeting; but during the time that had elapsed between the first moving of this business at the county of Longford meeting in June, and the provincial assemblies in September, the aspect of public affairs with regard to Ireland, and the position of the volunteers, had considerably changed. Peace was made with America, and with France. England had troops now at her disposal, and the volunteers ceased to be necessary for the defence of this country. Government no longer regarded them with a propitious and respectful, but rather with a jealous yet scornful eye; and were prepared to resist their further interference, and to watch their proceedings with a view to find reasons for their dissolution.

Attempts had already been made during the summer of 1782, to substitute what were called fencible regiments instead of the volunteer corps: the appointment of officers and their pay had been held out as a temptation by government, which some of the volunteers, it was whispered, would not be able to resist; but the measure was altogether so unpopular and so odious to the patriots, and to the great body of the volunteers, that those, who had been inclined to desert the popular cause, were anxious to keep their own secret, and the scheme of fencible regiments was abandoned. The disposition of government, however, was manifest, and gave sufficient warning to the volunteer patriots to be on their guard, both for their own sake, and for the good of the cause in which they were engaged. The reasonable and well-informed among them were sensible, that they had perhaps already snatched a grace beyond the reach of the constitution; nor did the applause, by which these happy efforts of patriotism had been crowned, prevent them from perceiving, that it must now be their care to defy criticism by a more strict conformity to constitutional rules.

That such was my father's opinion, given before the event justified the judgment, appears from the following passage in a letter to his most intimate friend, written a few days previous to the Leinster provincial assembly, where the question was to be discussed, whether the delegates for the future convention were in fact to be military from volunteer corps, or deputies from counties, constitutionally meeting to address and petition the parliament and the throne.

"I hear that the meeting on Thursday is to be very numerous. I am afraid I shall become unpopular, and be much attacked for endeavoring to persuade this assembly, not to chuse delegates for the national convention, but *to let the counties chuse for themselves.* But I think it my duty, to speak my opinion; and I know, that the mode of proceeding they propose would afterwards be an objection in parliament."

The scheme of sending delegates from volunteer corps was persisted in; but the meeting was more temperate in other points, than he had expected: their proceedings, after much debate, ended in a judicious resolution, that a reform should be sought *through the intervention of parliament.*

At length the convention, consisting of one hundred and sixty delegates from the volunteer corps, met at the Royal Exchange in Dublin, November the 9th, 1783.—Parliament was then sitting. An armed convention assembled in the capital, and sitting at the same time with the houses of Lords and Commons, deliberating on a legislative question, was a new and unprecedented spectacle. It was feared, from the general fermentation of men's minds, and from the particular enthusiasm of some of the delegates, that the convention might be hurried into acts of imprudence, and that affairs might not terminate happily.

In this convention, as in all public assemblies, there was a violent and a moderate party. Lord Charlemont, the president of the assembly, was at the head of the moderate men. Though not convinced of the strict legality of the meeting, he thought a reform in parliament so important and desirable an object, that to the probability or chance of obtaining this great advantage it was the wisdom of a true patriot to sacrifice punctilio, and to hazard all, but, what he was too wise and good to endanger, the peace of the country. Lord Charlemont accepted the office of president, specially with the hope, that he and his friends might be able to influence the convention in favor of proceedings at once temperate and firm. The very sincerity of his desire to attain a reform, rendered him clearsighted as to the means to be pursued; and while he wished that the people should be allowed every degree of liberty consistent with safety, no man was less inclined to democracy, or could feel more horror at the idea of involving his country in a state of civil anarchy.

The Bishop of Derry (Lord Bristol) wishing well to Ireland, but of a far less judicious character than Lord Charlemont, was at the head of the opposite party. Both were my father's private friends, and, upon this occasion, both honored him with their confidence. He had, therefore, as a private gentleman and a public man, a difficult course to pursue. Lord Charlemont, foreseeing the danger of disagreement between the parliament and convention, if at this time any communication were opened between them, earnestly deprecated the attempt. It was his desire, that the convention, after declaring their opinion in favor of a parliamentary reform, should adjourn without adopting a specific plan; and that they should refer it to future meetings of each county, to send to parliament, in the regular constitutional manner, their petitions and addresses. Mr. Flood, however, whose abilities and eloquence had predominant influence over the convention, and who wished to distinguish himself in parliament as the proposer of reform, prevailed upon the convention, on one of the last nights of their meeting, to send him, accompanied by other members of parliament from among the volunteer delegates, directly to the House of Commons then sitting. There he was to make a motion on the question of parliamentary reform, introducing to the House his specific plan from the convention. The appearance of Mr. Flood, and of the delegates by whom he was accompanied, in their volunteer uniforms, in the Irish House of Commons, excited an extraordinary sensation. Those who were present, and who have given an account of the scene that ensued, describe it as violent and tumultuous in the extreme. On both sides the passions were worked up to a dangerous height. The debate lasted all night. "The *tempest*, for, towards morning, *debate* there was none, at last ceased." The question was put, and Mr. Flood's motion for reform in parliament was negatived by a very large majority. The House of Commons then entered into resolutions declaratory of their fixed determination to maintain their just rights and privileges, against any encroachments whatever:—adding *that it was at that time indispensably necessary to make such a declaration.* Further, an address was moved, intended to be made the joint address of Lords and Commons to the throne, expressing their satisfaction with his Majesty's government, and their resolution to support that government, and the constitution, *with their lives and fortunes.* The address was carried up to the Lords, and immediately agreed to. This was done with the celerity of passion on all sides.

Meantime an armed convention continued sitting the whole night, waiting for the return of their delegates from the House of Commons, and impatient to learn the fate of Mr. Flood's motion. One step more, and irreparable, fatal imprudence might have been committed. Lord Charlemont, the president of the convention, felt the danger; and it required all the influence of his character, all the assistance of the friends of moderation, to prevail upon the assembly to dissolve, without waiting longer to hear the report from their delegates in the House of Commons. The convention had, in fact, nothing more to do, or nothing that they could attempt, without peril; but it was difficult to persuade the assembly to dissolve the meeting, and to return quietly to their respective counties and homes. This point, however, was fortunately accomplished, and early in the morning the meeting terminated.

I have heard my father say, that he ever afterwards rejoiced in the share he had in preserving one of the chiefs of this volunteer convention from a desperate resolution, and in determining the assembly to a temperate termination.

The following pages were written by him (in 1817,) in the last year of his life.

"I have been often urged by my family, to write an account of the political part of my life; but in fact it was during but a short period that I was ever engaged in politics. Twice only during my long life have I taken any active part in those great objects, which elicit the talents, and agitate the passions of mankind. The two great subjects of national discussion, which arose during my residence in Ireland, that called upon me to take a part in the politics of that country, were the attempt to obtain a parliamentary reform in 1783, and the question of a legislative union between Great Britain and Ireland in 1798.

" The history of the conventions, which were assembled for the purpose of obtaining a parliamentary reform, have been very imperfectly detailed in publications of no great merit, which appeared about that time when the business was an object of much party interest. In Lord Charlemont's life, written by Mr. Hardy, such circumstances as relate more immediately to Lord Charlemont are inserted; but Mr. Hardy had not the means of intimate or of impartial information. He was a man of retired habits; he was not a member of the convention; and he was particularly connected with friends, who were strongly interested in the continuance of that system, which gave patronage to the possessors of such boroughs as he represented in parliament. Lord Charlemont honored me with something more than common intimacy. He had appointed me to be one of his aides-de-camp, when he acted as commander in chief of the volunteer army of Ireland: besides, he honored me with his private confidence, so that I had the means of knowing his objects and his wishes, when he presided over the convention in Ireland in the year 1782.

" I was also well acquainted with the late Lord Bristol, Bishop of Derry, another of the leading men in that assembly. It happened, that I had known his Lordship on the continent, before the time of which I am now speaking. To know precisely the views of this extraordinary person was beyond my reach, and indeed I believe, that they were not accurately known even to himself. Popularity with his own party, and in his own neighbourhood, and the eager desire to shew his power to perplex and to obstruct the party which he opposed, were among the motives that actuated his conduct: with these inferior considerations there was certainly a desire to improve the defective representation of the people. It is not my intention to discuss the practicability or prudence of the measures then proposed; I wish to do no more, than to leave to my family some account of my own conduct when I was a member of the convention, and also of the changes, which time, new facts, and further experience, have effected upon my opinions as to the great political problem of a reform in parliament—a problem which has exercised and divided the opinions of the wisest and the best men in the empire. From a cursory view of the statements, which were made to the convention of the parliamentary representation in Ireland in the year 1782, it appeared, and it was universally believed, that there was an enormous monopoly of power, enjoyed by a small number of individuals, in the management of public affairs, and in the disposal of preferments. This, and the great influence in the appropriation of the public money, could not fail to raise indignation in the minds of those, who felt themselves excluded from a participation in the advantages, which were thus exclusively consigned to a few, whose talents and virtues were not always superior to those of the generality of their fellow-citizens. This had excited a strong, and something like a republican desire, to reduce the members of the oligarchy of Ireland to a level with themselves; and the ministers of the crown in England were not sorry to find some inroads made upon that confederacy, which had long, and in no small degree, constrained the authority of the Castle. Perhaps no assembly in any country ever contained more of the proprietors of the soil of the country, or of men of rank, influence, information, and real patriotism, than were collected upon this extraordinary occasion.

" The 10th of November, 1783, the volunteer convention assembled at the Rotunda in Dublin; and I venture to affirm, that there never was any assembly in the British empire more in earnest in the business on which they were convened, or less influenced by courtly interference or cabal. But the object was in itself unattainable.

The idea of admitting the Roman Catholics to the right of voting for representatives was not urged even by the most liberal and most enlightened members of the convention; and the number, and wealth, and knowledge of Protestant voters in Ireland, could not decently be considered as sufficient, to elect an adequate and fair representation of the people.

With respect to the question of parliamentary reform, it can scarcely be doubted, that some change in the borough system appeared desirable to men of the coolest judgment and of the soundest principles; but I am ready to confess, that my own opinions then went farther than I can now (in 1817) approve. The idea of universal suffrage and of annual parliaments never was entertained by any reasonable person of that convention; but some of the best wishers to the constitution were led farther away from the influence of aristocracy than was prudent. The proceedings of the convention are in print, and the speeches of its members, though very ill reported, give some idea of the sentiments and views of the different parties, of which it was composed.

" I shall, however, mention a circumstance, which has not, I believe, been generally known. On one of the last days of the convention, the Bishop of Derry (Lord Bristol) invited the most active of the members to a dinner. Before the bottle had made any considerable impression upon the company, it was proposed, that a motion should be made in the convention, for carrying up its petition to the door of the House of Commons, by the whole convention in their uniforms!

"I was the first person who opposed this proposition; and, fortunately for the country, all present became convinced of its rashness and illegality. I assert, positively, that *I was the person, who ventured, in a distinct manner, first to oppose this proposition.* I was well satisfied at the time, and I am now certain, that this was a plan, which the Bishop of Derry was eager to promote; and that the company then selected, and the dinner then given, were prepared for the express purpose of collecting a party, that would forward the measure in the convention. This dinner was given at the house of the nephew of the Bishop of Derry, Mr. George Robert Fitzgerald, of famous memory. It is a circumstance worth remarking, that Mr. Grattan and his lady dined with us, but retired very soon after dinner: this was managed by the Bishop, to give an importance to the meeting; for, had Mr. Grattan remained to hear the plan that was proposed, he would have reprobated the scheme with contempt and indignation.

Some time after this circumstance took place, I heard more than one gentleman, who was of that company, claim the merit of having prevented the convention from going in arms to the House of Commons; and, what is extraordinary, I do believe, that each of these gentlemen was under some degree of self-delusion. I have now, however, before me a letter from the Bishop of Derry, written soon after the dissolution of the convention, reproaching *me* for having counteracted his wishes, and having rendered his plan abortive.

"Upon the whole, the rise, progress, and dissolution, of this extraordinary assembly, and of the volunteer army of

Ireland, is a most singular political phenomenon, and will remain a lasting memorial of the good sense and loyalty of the gentlemen of Ireland.

"To cut off the rotten Irish boroughs, and to substitute members chosen from rich and populous places in their stead, was a measure so consonant to common sense and common justice, as to require for its support neither oratory nor numbers. To remove an obvious cause of reproach, inconsistency, and criminality, from the legislature, and thus to uphold its dignity and justice in the eyes of the people—was, in itself, an object well worthy the interference of true patriots; and would, if it had been adopted by the government of the country, have added force and real influence to the administration, without disturbing the economy of the constitution, or weakening the just power of the aristocracy."

After the termination of the convention, the hopes of those who desired a Parliamentary Reform were not abandoned; and these revived, at times, with fresh energy, during the years 1784 and 1785. Soon after the peace, there was a change of ministry. Mr. Pitt, who, in his early days, while in opposition, had been one of the great advocates in England for Parliamentary Reform, became minister; and it was confidently expected by the Irish patriots, that he would favor their efforts. Fresh meetings, congresses, and conventions, were held, at various times and places. During the Duke of Rutland's vice-regency much was expected. But the result of the whole was, that nothing was done. At one time, the Lord-Lieutenant said, he would transmit, as he was requested, certain Dungannon addresses; but that he could not do so without marking his disapprobation of the unjust reflections thrown by them on his Majesty's parliament. At another time, the sheriffs of Dublin were fined by government, for calling or permitting a new meeting of volunteer delegates, for the purpose of pursuing this object of reform. Mr. Flood was repeatedly foiled in his attempts to bring forward the measure in parliament. While there was any reasonable expectation of success, or while the object was pursued by any reasonable means, my father, convinced as he then was of its being a measure advantageous to his country, pursued it with uniform consistency, and zeal. At length, the hopes of the reformists died away, and the attention of the volunteers and patriots turned to other objects.

From this time, till he went into the

Irish parliament, many years afterwards, he took no further part in public affairs. And now, quitting a subject to which I am unused and unequal, I return, with pleasure, from politics to domestic life.

CHAPTER IV.

[1782—1789.

NEXT to biography written by the person himself, his private letters afford the best means of obtaining an insight into his character. But some readers, and those whose good opinion is most valuable, those of the most honorable minds, will recoil at the idea of publishing letters.—They need not here apprehend any breach of private confidence.—No one could have a greater horror than my father felt at the publication of private letters, such as have appeared in too many instances, unworthy of the characters of the illustrious dead, painful to the living, injurious to surviving families, and fit merely to gratify idle, or, worse than idle, malignant curiosity. With respect to the affairs of others, and to any confidence ever reposed in him, my father was too honorably exact to leave to chance, or to transfer to the discretion of any other person, that which had been exclusively entrusted to his own.

Two years before his death, he burned some thousands of letters, many of them most entertaining, and from persons of literary celebrity. He permitted Mrs. Edgeworth to snatch some of the late Dr. Darwin's from the flames; and she proved to him, upon looking these over, that no possible harm could happen from their being preserved; but he persisted in burning the rest. He said, he felt that he should not have time to look over all his friends' letters, and that his mind could not feel at ease, unless those which he had not examined were destroyed. Many years previous to this time, Mr. Day had given him permission to make whatever use he thought proper of his letters; and some had been looked over for publication, when my father had thoughts of writing Mr. Day's life. Not one other letter—not a line in any letter, is here given to the public, by me, without permission from those surviving friends and near relations, who are the proper guardians of the fame of the dead.

With regard to his own letters, I should observe, that these are not published from *copies:* he kept copies only of letters of business. His original letters to Mr. Day and Dr. Darwin were returned to us after their deaths: some of his to the late Mrs. Edgeworth were also entrusted to me; and from these I have selected a few, which have not any pretension to literary value, but which may shew what were his habits of domestic life, and the course of his thoughts and occupations, during a long period of years. Without need of my assertions, every reader of discernment and feeling will at once perceive, that these letters are not *manufactured* for the public, but written in the fulness of the heart, and in the careless ease of private confidence. Such are the only letters worth seeing, yet how few of such are fit to meet the public eye! There can scarcely be a stronger proof of any man's consistency and single-heartedness, than that his best friends can dare to lay before the public his *really* private correspondence. Considered in this point of view, I trust, that, in producing my father's letters to various persons, at different and distant periods of his life, I shall do honour to his memory, with all impartial judges. But, if I err in publishing any of his letters, I earnestly supplicate, that the blame may rest where it ought to fall,—*on me alone.*

TO MRS. EDGEWORTH.

Dublin, Dec. 7, 1782.

"YES, my dear wife, I have the delightful prospect before me of having, in the place of my dearest Honora, another friend whom I love, and whose company, I hope, will make age a pleasing calm, after the activity of youth and the business of middle life. And this being my middle age, it is time to talk of business. I called on the upholsterer, and executed all your commissions.

"Nothing but extreme diffidence prevents you from being convinced of your abilities as much as I am. Pray, to try your modesty, tell Maria, that all your corrections but one are adopted in the two quires of the MS.*, which I have looked over; and that they are much more correct and better written than the former. I expect the remainder every day from Edgeworth-Town, and you may assure her, that, if I sit up all night to correct them, I will not delay their conveyance to England. Cecile's letter is much the

* A translation of Madame de Genlis' Adele et Théodore, ou Lettres sur l'Education.

worst done—It is *not well.* Chevalier Herbain's is the best. Most misses would have made some hand of the former, and none of the latter. I have not gone to see *any body* yet—nor one tenth finished my private business; though the weather has been particularly favourable, and has let me walk about town constantly from eleven to four. I have spent every evening but one at home, from dinner time to half-past twelve, chiefly at Maria's manuscript.

"I thought a sheet of paper would have gone a great way, or rather, would have let me go a great way, in telling you so short a piece of information, as that 'my affection and esteem for you increase every day.' For in fact I had little else to say, and beginning at the top of a first page was so much more enlarged a view than I have in common, that I begin to think I grow old, or, as Caractacus exclaims, 'Gods, I grow a prattler!' "Adieu."

In 1785, the Royal Irish Academy was formed from a literary society, that used to meet in Dublin every week to read papers on philosophical subjects. It became an object with many Irish gentlemen, to be of the number of original members: my father's friend, the late ingenious Doctor Usher, then professor of astronomy in the University of Dublin, proposed him; but I find by the following memorandum, that he owed the insertion of his name in the patent, at the very last moment, to Lord Charlemont, who, from the time of the Irish Convention, had honoured him with peculiar regard.

"When an application was made to the Crown for the establishment of the Royal Irish Academy, a small number of gentlemen were named as its original members.

"Several applied ineffectually to be admitted on this list; but upon my desiring to have the honor of being enrolled as one of the founders of the Royal Irish Academy, his Lordship prevailed upon his associates to have the patent opened, and to insert my name in it, just as they were going to send it over to England for the royal sanction."

There was a mistake in the manner in which his name was entered in the patent. It was written "Lovell Edgeworth," instead of Richard Lovell Edgeworth. This is worth mentioning, because it has led to a confusion of persons in various publications, in which my father's name and my brother's have been quoted.

During the years 1786 and 1787, my father was occupied with mechanics and agriculture. He improved, with success and profit, considerable tracts of mountainous land, and of bog. In carrying limestone for the improvement of the mountain farm, he made the first trial of wooden moveable railways, and small carriages with cast-iron wheels, supported on friction rollers; as I find by the following passages in his letters he proposed, as early as the year 1786, to employ these moveable railways in public works in Ireland.

TO MRS. EDGEWORTH.

"*Dublin, Feb.* 1786.

"I GAIN ground over the incredulity of the good folks of this *great* capital, and they now begin to think it possible, to carry goods by a cheaper conveyance than an Irish car.

"I make no doubt, if I apply for a clause in an act of parliament, to permit me to lay railways, that I shall succeed. On all hands I am told, that my proposals are the fairest and most disinterested, that have been made by any man, or any set of men, in Ireland; and, indeed, I think they are.

"I attribute, my dear Bessy, your anxiety lest I should engage too ardently in this business to that true friendship and affection, which you invariably shew me upon every occasion of consequence; and I can only reply, that one word from you can at any time abate my application, or, if necessary, stop me in the midst of my career.

"R. L. E."

TO THE SAME.

"*Dublin, 25th Feb.* 1786.

"A PRIORI I decided, that it would be impossible, that I should be able to write to you to night; but experience always teaches me, that I *do* find time to write, if it be but one line, to my dearest and kindest friend.

" I have seen gentlemen of all sorts, upon my business, since I came to town, and have been much surprised at the extreme ignorance, which some of them have shewn, upon subjects where they are immediately concerned.

" No specific demand for coals is yet determined, but I think in a few days I shall be able to ascertain what may be done.

" I saw the Speaker this evening at the house, and mentioned my idea of applying for a clause in the Navigation Act, to permit me to lay railways upon the banks of unfinished canals.

" He was particularly kind, and immediately mentioned it to several members with approbation; taking the opportunity to relate the success of the experiment on this subject, which I tried at his own house last summer.

" R. L. E."

TO THE SAME.

" *Dublin, March* 30, 1786.

" AMONGST the numerous proofs that you have given, and continue to give me, my dearest Bessy, of your confidence and affection, none strike me with more satisfaction, than the ease and spirit with which you write to me. A ray of health gleamed through your last, that gave me infinite pleasure; and the content and moderation, which you continue to express, gave me the most solid security for lasting happiness. I am fully persuaded, as much so as you can be, that more wealth might add care, but could scarcely make an addition of real enjoyment to our present situation.

"I believe, also, that I know how to estimate the transient breath of fame. But I never thought it wise, to repress what little enthusiasm might remain, after what I have seen of life. I know, that without it I must necessarily cease to act; and laborious employment, tending by degrees to some object of our wish, I firmly believe to be amongst the chief pleasures of existence.

"I doubt whether I ever wrote to you so much without a word of business. I will tell you, therefore, in three words—that I have finished and shewn a machine to the Speaker, that carries near four tons, and is moved by one man, and that I am taking out a patent. I have a prospect of a short, easy project in the county of Kildare, to try my plan upon, that will do me no great hurt if it fails. But I propose myself more happiness in returning home on friday, than any project can bestow on your affectionate

"R. L. E."

TO DR. DARWIN.

"MY DEAR FRIEND AND DOCTOR,

"I SET out to morrow, to conclude the preliminaries of a bargain with the undertakers of a great iron-work, which is going to be established in this neighbourhood, for the carriage of all their ore, coals, &c. The quantity above 20,000 tons—the distance less than a mile. For this purpose, I lay railways, and use certain carriages of my own construction upon friction-wheels, &c., upon the principle of having a number of small carriages linked together, instead of using large machines and expensive railways. By this contrivance, I am enabled to have slight moveable railways easily laid, and easily repaired, &c. &c.

"I have already applied these machines to the conveyance of lime-stone gravel, for improving ground, and have found them, from *two years* experience, (time enough to try their durability and wear and tear) highly profitable. When I have concluded my preliminaries, I will tell you the terms. Another more useful scheme that I have in view is, to carry marl two miles with this machinery, to improve 600 acres of coarse land, which has just, by the extinction of leases, fallen into my own hands. Such parts of it as have been limed pay 15*s.* per acre; unimproved, 3*s.* I hope to marl 200 acres next summer, at the expense of 10*s.* each; that is to say, if you will set my health up again.

* * * * *

Fortunately before he engaged with the company, for whom he had prepared to lay his railways, he discovered, that they had then little to carry, and nothing to pay. By the use he made of this machinery on his own estate, he was recompensed for his trouble; and the success of his experiments in these first trials convinced him, that they would lead to future extensive benefit to the public.

The next letters to Dr. Darwin speak of his having had a dangerous attack of illness. As he was standing upon a scaffold, giving directions to some masons, he was seized with a giddiness in his head, so sudden and violent, that he fell from the scaffold into an area beneath, many feet deep. The workmen thought that he was killed. Two of them, carrying him between them, brought him into the house. In a few moments, he relieved us from terror, by speaking in a cheerful tone and playful manner. He told us of a French marquis, who fell from a balcony, at Versailles, and who, as it was court politeness, that nothing unfortunate should ever be mentioned in the king's presence, replied to his majesty's inquiry, if he was hurt by his fall, "Tout au contraire, Sire." To all our inquiries, whether he was hurt, my father replied, "Tout au contraire, mes amies." He assured us, that he thought the shock had been of use to his nerves,—that it would teach him to mind better what he was about, and to take better care of himself.

His cheerfulness quieted our fears for the time; but he did not recover his health, in consequence of this *salutary* shock to his nerves. Dr. Darwin predicted, that the disease would terminate in jaundice.

Meantime, though under the increasing pressure of a disorder, which is said to dispose the patient to be peculiarly inactive and desponding, he continued his usual employments and amusements. We were just then preparing to act a little play; he was fitting up a theatre for us, and contriving a new manner in which the side scenes should turn on pivots, instead of sliding in the common method. But he was taken so ill, that he was forced to stop; and the next morning, the jaundice, in dreadful yellow, appeared. The following letter to Dr. Darwin shews, however, that jaundice did not make him fretful or melancholy:—

TO DR. DARWIN.

" *Edgeworth-Town*, 1786.

" MY DEAR DOCTOR,

" YOU used to laugh at me for my fondness of the 'Arabian Tales;' but I have always found them a mine of inexhaustible treasure.

" Three brothers, in one of these tales, set out to discover some useful present for a princess, to whom they pay their court. One procures a telescope, which brings any object you wish, let its distance be ever so great, immediately before your eyes. The second, a sofa, which transports itself through the air with incredible velocity; viz. a balloon. And the third, a golden apple, the odour of which instantly restores the health of those who smell it.

" The three princes meet at a great distance from their mistress's country, and agree to shew each other their curiosities. The first looks through his telescope, and sees the lady at the point of death. The second transports himself and his brothers to the sick apartment, and the apple of the third revives her. They dispute the value of their treasures, each urging, that, without his own, the princess must have died, &c.

" You, my dear Doctor, did, by the telescopic eye of medical sagacity, perceive my disease, before it was visible to vulgar sight. The post wafteth your wisdom across the ocean with magical celerity. And I hope, that you will send me the Golden Apple, to complete the end of my story happily. * * * * * [*Details of his health omitted.*] ——I never took any drug, except of your prescription.— But my will is in the hand of my friend.——

" Pray teach me in time, for the spring advanceth, how to convert colourless water into green grass.—[*Dr. Darwin was, at this time, trying experiments on the irrigation of meadows.*]

" Thank you again and again for your kindness—and Mrs. E. and four daughters thank you also. We have just been acting a little farce, for our own family and intimate visitors *only*. The piece written, and all the characters filled, by ourselves.

" We promised Lord and Lady Longford and their children, who came to stay some days with us, that we would give them a play of *home-manufacture*. I send you our prologue. You recollect, I am sure, Mrs. Greville's

" Prayer for Indifference," addressed to Oberon. Mrs. Greville's family and the Pakenhams are connected: and I must further instruct you, that the river Inny separates Lord Longford from us: now you will understand this little prologue.

" *Iterum, iterumque vale,*

" R. L. E."

" *Prologue, in the character of Oberon, spoken by Charlotte E. (between three and four years old).*

" Deaf to accomplish'd Greville's hasty prayer,
I hurl my spells into the spungy air;
Oberon my name, of magic skill, to raise
The drooping spirits, by the power of praise;
Indifference with this fairy wand I charm;
With this, each youthful, friendly, bosom warm;
So shall the mimic visions of the night
Delude your judgment, and enchant your sight;
And lords and ladies, who, for Christmas gambol,
To see this farce, across the Inny ramble,
Dull though it be, shall never cease to smile,
Through three long acts shall sit the tedious while;
Nor dare to hiss, but clap each little actor,
Content, though coarse, with home-made manufacture."

I am sorry that Dr. Darwin's next letter was burned; I regret it not for the sake of compliments to play or prologue, but for the sake of a humourous invective, which it contained against what the Doctor called *vinous potation.* He believed, that almost all the distempers of the higher classes of people arise from drinking, in some form or other, too much vinous spirit. To this he attributed the aristocratic disease of gout, the jaundice, and all bilious or liver complaints; in short, all the family of pain. This opinion he supported in his writings with the force of his eloquence and reason; and still more in conversation, by all those powers of wit, satire, and peculiar humor, which never appeared fully to the public in his works, but which gained him strong ascendancy in private society. During his life-time, he almost banished wine from the tables of the rich of his acquaintance; and persuaded most of the gentry in his own and the neighbouring counties, to become water-drinkers. Partly in jest and partly in earnest, he expressed his suspicions, and carried his inferences on this subject, to a preposterous excess. When he heard that my father was bilious, he suspected, that this must be the consequence of his having, since his residence in Ireland, and in compliance with the fashion of the country, indulged too freely in drinking. His letter,

I remember, concluded with—" Farewell, " my dear friend. God keep you from " whiskey—if he can."

To any one who knew my father, this must seem a laughable suspicion, for he was famous, or in those days, I may say, infamous, in Ireland, for his temperance.

TO DR. DARWIN.

" *Edgeworth-Town*, *Oct.* 23, 1786.

" We were all highly entertained and delighted with your letter. Mrs. Edgeworth, because it relieved her from some apprehension; Miss E. because it was witty; and I, my dear Doctor, because it was wise. I cannot, however, help asking why the entire class of middling people in this country, who drink intemperately of whiskey-punch, are exempt from the gout, gravel, and stone. They are not often dropsical, and, for the most part, they live to an old age. Particular circumstances make me acquainted with this class of people, and, from many years experience, I assert, that they are, in the proportion of twenty to one, exempt from these diseases; and that there is not one, out of twenty, who does not drink *strong* vinous spirit. Many of them regularly take a glass, every morning, of spirits as strong as the best rum.

" All the lower class drink as much whiskey as they can get, and are, notwithstanding, strong and healthy!

" The gentlemen who drink wine, and eat luxuriously, are, on the contrary, afflicted with all the demons of disease, and flock to Bath and Spa, like regular birds of passage, every autumn.

" I have had three returns of my complaint * * * * * I have been surprised to find, that passing several sleepless nights in pain has not affected me on the succeeding days nearly so much, as want of rest has done when I have been in better health. I shall, as soon as I am stronger, come down from three (rarely four) to one or two glasses of wine; indeed, when I am well, I seldom exceed two, FOR I love not that which you call vinous potation. I am as *notorious* for sobriety here, as *you* are in England.

" I wish I had obeyed your orders, and written on larger paper: for I want to tell you of the amazing profit obtained by the improvement of land in Ireland. Your procuring me an account of the best method of floating meadows shall be repaid by an accurate account of its success in this country. I suspect, that it depends chiefly upon the soil through which the water passes: the improvement of the land adjoining to peat-mosses, or, as we call them, turf-bogs, is the most profitable I have ever attempted. I think, where the ground is chosen with judgment, it brings in three years cent per cent.

" A ditch canal would not do as well as *my* railways, for the manure must be carried sometimes up hill. A fire-balloon I had thought of, but upon calculation you will find, that it would not succeed. I congratulate you upon my friend, Dr. Robert* Darwin, being settled. I make no doubt of his abilities. A friend of mine, Mr. William Brooke, (now Dr. Brooke of Dublin) is gone to Edinburgh, where one word from you will serve him. His good sense and good conduct I will answer for.

" R. L. E."

* Now Dr. Darwin of Shrewsbury.

FROM MR. DAY TO MR. EDGEWORTH.

1787.

* * * * * * * * * "After all I must confess, that my experience for some years past has made me look with great concern upon the perishable nature of human friendships. With by much the greater part of mankind the name means absolutely nothing; a traffic of interest or vanity; a league of knaves and fools—of sharpers and bubbles; a party of pleasure, made in the morning, and broken off, with disgust, at night; an agreement to travel in a post-chaise dissolved at the first inn by a squabble about the reckoning; a theatrical exhibition, where the two heroes strut about, and mouth their parts, and then retire to undress and abuse each other in the green-room. But the worst is, that even men of more ingenuous minds and liberal characters are not entirely exempt from these general laws. If neither interest nor any other passion intervenes, still there will be different points of view, from which they consider the same object; and dissimilar perceptions can scarcely be reduced to any common measure. Even men of this description are subject to their passions, though of a nobler nature; and these passions are perpetually busy to alter the colors and magnitude of every object which they consider.

"For this defect of human nature I absolutely know no remedy. All that my reason furnishes me with upon the subject is, to contract the circle of my own wants and cares, and to trouble others as little as possible for any thing but strict justice; and even this is a commodity not to be easily had. In respect to those who honor me with their friendship, I study to make myself worthy of it, by such a kind of conduct as will never render their connexions with me a subject of disgrace or repentance.

"Whatever may be my precise opinions, I am generally ready, upon the most disinterested principles, to do for them, at least as much as they can pretend to have experienced from others. If, after all, I am misunderstood, or misunderstand my own character, I am, I fear, too old to mend; because the fault must be in my original and simple perception of things, which cannot be rectified; and nothing remains for me, excepting to shrug my shoulders, with Brydone's poor Italians, and to cry 'Pazienza.'

"I am, with the greatest sincerity,

"Your affectionate friend,

"T. Day."

The rhetorical style of this, and of Mr. Day's letters in general, may, perhaps, give the idea of their being composed with care. On the contrary, he wrote them as fast as his pen could move: his usual manner in speaking resembled these letters. As the common people expressed it, Mr. Day always *talked like a book*—and I do believe he always thought in the same full-dress style. This was the result of the systematic care he had early taken, to make himself master of his native language, and to cultivate eloquence.

FROM MR. DAY.

" MY DEAR FRIEND, 1788.

" I MUST confess, that you are very right in saying, that I am always armed, and ready to repel aggression. I am conscious, that there is some truth in the charge, and the daily observations I make in the world do not tend to abate the disposition.

" I never expected any thing very romantic from my fellow creatures, and have long confined myself to the lowest returns of good behaviour, or the commonest attentions of civility. Yet daily experience shews, that even this is too much to be expected from those, who make the greatest professions, and have most experienced your friendship. So wonderfully do events turn up, that I really am surprised, that I am not become quite misanthropical, and alienated from the species. But this is certainly not the case, as I never was better humored in my life, or more inclined to serve them, within certain moderate bounds.

" Much of this certainly arises from our forming our connexions rather from accident than judgment; and taking the next, rather than the most approved, as a subject for our social affections.

" I never was much inclined to believe in the agency of the Devil, nor do I now attribute so much to the malignity as to the folly of mankind.

" The general characteristic of our species is certainly levity and inconstancy. When they make the greatest quantity of professions, they frequently do not know themselves, that they neither feel, nor will perform, one thousandth part of what they promise; and when they appear to act with the greatest degree of ingratitude, they are generally only treating you as they do their most essential interests:—Seeing men in this point of view, it is certainly very easy to forgive them, but not very easy to attribute much consistency, or dignity, to their conduct.

"As to yourself, you very well know, I have always considered your conduct in the most disinterested light, and your attentions to me, as solely founded upon personal esteem and benevolence; therefore any of these reflections cannot, either directly or indirectly, carry with them any sarcastical inuendo. As to the characters of my particular friends, I cannot admit, or believe, that I have ever been much deceived. * * * * * *

"But men, as they advance in life, pursue dissimilar studies, and contract incongruous habits, which render them less fit for each other's society. Thus, a gradual strangeness takes place, without one party being more blameable than another. Since our gratitude must depend not upon our judgment but our habitual intimacy,—upon habitual feelings of pain or pleasure; and it is as impossible to like a certain set of manners, as it is to like a certain set of features—when they are habitually displeasing.

"In this case, I think nothing more natural and rational, than for people to pursue their own plans and amusements at a distance, and to slip their necks out of the collar, when they do not relish the weight. * * * * * * * * . I own, that my observations have made me depend much more upon my own judgment than I used to do; because, weak and fallible as I must confess this to be, I am not sure of meeting with more impartiality elsewhere. I may, indeed, be totally wrong in all my judgments, but this opinion would prevent my acting at all. Nor could I have any better reason for believing implicitly in the judgment of another, than in my own.

" I cannot help digressing here, to propose a curious argument, derived from this principle, against the church of Rome—which I do not remember to have seen. He that asserts the infallibility of another, must also assert his own; otherwise he may be deceived in the judgment he makes of that infallibility, as well as in any other judgment. But if he allow, that all his own judgments are fallible, and may be erroneous, then his particular opinion of the infallibility may be erroneous too, unless he can shew a particular reason for the exception. In this manner it may be shewn, that the real confidence every man has in his own judgment is much the same, since it must always precede his having a confidence in any one else.

" Like Montaigne, I have almost written my essay without coming to the subject of it. * * * * * * * *

" Sincerely and affectionately yours,

" T. Day."

The subject, to which he alludes, related to private affairs, and therefore the conclusion of this letter is omitted. The general opinion he gave was, that generosity is more frequently injurious than beneficial, both to those who practise it, and to those who are its object. He had, by this time, brought himself to think, that assisting or promising to assist friends, who have been imprudent, only encourages extravagance or carelessness, and excites unreasonable expectations. My father, struck with the great change which appeared in Mr. Day's sentiments, compared with those of his youth, wrote to him as follows:

FROM MR. EDGEWORTH TO MR. DAY.

"MY DEAR FRIEND, 1788.

"* * * You say, with great truth, that it is much wiser to withhold our promises, and to do whatever good we can from the impulse of occurrences, rather than from the obligation of a contract. * * * * * * * * As to the general question of what one should do for one's friends in any given situation, every thing relative to supererogatory duties must depend upon the feelings of the person, who thinks himself called upon, and not upon the judgment of indifferent spectators. These feelings must vary infinitely, as well from the degree of intimacy, as from the former experience of each individual. I neither wish to lessen you, nor to pay myself a compliment, in supposing, that we may be naturally equally disposed to generosity; but during the course of our lives, our manner of exerting this quality has been different. You had early in life the command of money; and, from your very childhood, you were certain of having a considerable fortune to dispose of, when you should come of age. Your school-fellows were not in the same situation, and you found premature occasions for bestowing favors amongst persons, not your equals either in fortune or benevolence. This continued through a large portion of your life. The wants of some, and the ingratitude of others, disgusted you, and taught you at last to be more circumspect in future. I suppose you have given away, at least, ten times as much as I have done; and I believe, that you have met with *twenty* times more ingratitude. Our experience, therefore, leads us necessarily to different conclusions. I think, that, perhaps, twice out of thrice I shall be ill treated. You think, that you will be ill used nine times out of ten.

" Another circumstance leads me to more confidence than you are disposed to. I have generally found means to execute whatever I have engaged to do; and I am, therefore, more inclined to form engagements, than, I believe, on cool reflection, prudence would permit. I have also habituated my vanity to be more pleased with executing an engagement, than with doing an extempore favor. I feel my enthusiasm increase from the moment I begin what I have once engaged to execute. I can trace the cause of this feeling to my mother's precepts. She early warned me of the violence of my temper, and of my passions; and, with true prudence, excited my ambition to a steady performance of whatever I might begin.

" You apologize for a desultory letter: What must I do ?—though you are certainly the ablest critic, and by far the best writer of my acquaintance, I never could bring myself to think of the manner in which I write to you. My opinions always produce themselves before you, without any exertion of my own. A certain sense of equality and independence has always prevented me from yielding to your judgment on one hand, or opposing it, with obstinacy, on the other. " Adieu.—Yours,

" R. L. E."

FROM MR. DAY.

Anningsley, 1789.

* * * * * * * " Relative to the subject of farming, I should almost look upon your desire for information as a boast of superiority: for I think you must be sufficiently

acquainted with the principle of it in England, to know, that gentlemen are much more likely to lose, than to gain 40 per cent.—I will however give you the detail you desire, with all the precision possible.

In the first place then, I am out of pocket every year about 300*l*. by the farm I keep. The soil I have taken in hand, I am convinced, is one of the most completely barren in England. The estate is certainly improved in value by what I have done; but were it to be let, I do not imagine that it would pay 2 per cent for all that I have laid out, probably not above one. You may perhaps wonder, I should persevere in such a losing trade; and, to avoid future explanation, I will give it you now. I am particularly pleased with the study of agriculture; and the constant business, which so large a farm creates, gives me the most agreeable interest in the world.—It gives me a continual object in going out: and the necessary trouble of governing so many men, and providing for so many animals, keeps my mind from stagnating in solitude. By these means I am enabled to live happily, with a perfect independence of my fellow creatures; for the succession of employments is such, that my whole life is taken up, without fatigue, or ennui.

Were I to give up farming, I should have less care, but I should also become more sedentary; and the very absence of that care, which now never rises to any thing like uneasiness, would expose me infinitely more to hypochondriacism, which I am now totally free from. I have besides another very material reason, which is, that it enables me to employ the poor; and the result of all my speculations about humanity is, that the only way of benefiting mankind is to give them employment, and make them *earn* their money. Besides this, I have even a very substantial motive of interest: I do not want to practise the *obliquity* of great accumulations, but I now think, that every wise man will rather improve his circumstances, as it is almost impossible not either to improve or to diminish. I take care to live within my income, and while I spend large sums in employing the poor, they are not entirely lost. I am continually improving the quality of my lands, and the conveniences of my estate. Buildings, planting, every kind of improvement, I chuse to pay for out of my income. I consider the pleasure of every thing to lie in the pursuit; and therefore, while I am contented with the conveniences I enjoy, it is a matter of indifference whether I am five or twenty years in completing my intended plans. This scheme also is connected with my own particular temper; for doing nothing with relation to the opinion of others, and every thing from a thorough knowledge of my own tastes and feelings, I do nothing that does not permanently please me.

There is also another very ample source of future profit, in the very large plantations I am continually making, and which in no very long space of time will probably more than double the value of my estates. Nor are these improvements confined to the spot where I live: I have purchased about 500 acres of indifferent land in the neighbourhood, the greater part of which I am covering with plantations. Most of the farmers, indeed all, are by particular circumstances extremely dependent upon me. I have lost something by the acquisition of this dependence, for you may easily conjecture, that the price of it has been uncommon forbearance in respect to their rents. But I do not know, that I have lost more than I should have done by turning them out of their farms. Some of them are honest, industrious men, who pay their rents well, but are near the expiration of their leases. These I continue upon their farm without any addition to their rent—with the condition only of letting me yearly plant a small quantity of their farms. By these means, I have not a farm, whose value will not in time be doubled or trebled. I have purchased all these within the last six years, and have taken care to make purchases, where I could have a large extent of land for little money; by all these precautions the general value of my property is increasing yearly. Were they to be sold, all these plantations would much enhance their value; and should they not change hands, it will be many years before my first plantation will require thinning; and from that era there will be a certain produce, increasing yearly in value, till in time it would amount, I should suppose, to more than the income of my present fortune. * * * * * * * * And now having given you, I think, all the information you require, I shall beg leave to rest my wearied pen, and subscribe myself

" Affectionately yours,

" T. Day."

I cannot help observing upon this letter, that Mr. Day differs from many philosophers in one remarkable particular, in being more benevolent in practice than in theory. While he fancies himself a misanthropist, he is exerting his time and faculties, and expending an ample fortune, for beneficent purposes, relieving to the utmost of his power all the wants of his fellow-creatures.

FROM MR. DAY TO MR. EDGEWORTH.

On a plan which Mr. E. had proposed of taking a child from the lower class of life to educate for a higher rank.

178—.

* * * * * * * "There is a very strong reason against it, contained in the following observations. In our state of society, almost every person is brought up to very expensive wants, and a very small degree of exertion to satisfy them; therefore, whenever they have the opportunity of doing it, they will very kindly transfer the whole burthen to another. But the eternal law of nature, and when I say nature, I always mean living intelligence, has determined, that the sum of what any given society can possess, is very little more than the amount of their laborious work.

"Says nature, 'Dig, plough, grub, fish, hunt, build, and you will be rewarded for your pains.'—'No,' say the French, the English, or some other refined people; 'we chuse to be idle and sentimental.'—'Then starve,' says nature—'this is my eternal, immutable decision, of which neither plays, nor poetry, nor oratory, nor sentiment, will ever change one tittle.'

"What then is the lot of man, in every country? To labor, and eat his bread with the sweat of his brow. If from the inevitable variety of combination in every civilized society there are a certain number, who are exceptions to the rule, nature has nothing to do with the subject; and it is equally indifferent to her Ladyship, which letter in the alphabet constitutes the dignified exception. What then is the proper employment of benevolence below? In my opinion, to rectify as far as we are able (and that is very little indeed) the evils, which proceed from the unequal distribution of property—by relieving those, first, who are in absolute want of the common necessaries of life, and particularly those who want them without their own fault. If we chuse to make a lady out of what fortune has intended for a serving wench, or a gentleman out of the materials of a blacksmith, we certainly have a very good right; but we err extremely, if we imagine we are either promoting the order of nature, or the good of society; for there will in every country be more than a sufficient crop of gentlemen and ladies, growing up like thistles among the corn. * * * * * * * * *

* * "In respect to taking another person's child to educate, it appears to me so serious a thing, that were it any indifferent person, I should suspect, that they had never considered the subject.

"That child, from the moment it begins to reflect, will consider you as doomed to supply all its wants, and will infallibly proportion his expectations, not to any standard of justice or reason, but to the objects that are presented to his eyes. Should he be idle, should he be vicious, will you continue to supply all his wants, and maintain as a gentleman him whom you have taken as a beggar? Or will you much relish, towards the decline of life, the having manifestoes to publish about your own conduct, and to apologise to your fellow-creatures for not being a dupe, or an idiot? Those fellow-creatures, believe me, will always be glad to abuse those whom they respect, or dread; and rejoice exceedingly to show the goodness of their own hearts, without any other trouble or expense, than that of abusing their neighbours. Even if he turns out well, will you support him as a gentleman? or buy him a commission in the army, a living in the church, or allow him three hundred a year, while he is waiting for business at the bar? This is upon the supposition, that he is to be bred a gentleman. If to any mechanical business and the use of his hands, I should think it better for him, to be accustomed to less splendid objects, and more humble hopes, than your family, moderate as it is, will ever present. The success of human beings is never proportional to their advantages, but to their exertions.

"If such a child, therefore, is to work for his own living, are you not likely to do him more hurt, by taking from him stronger motives to exertion, than good, by bestowing upon him as many barrels of potatoes, as many kilderkins of milk, as will nourish him till he is sixteen? Nor is it necessary, if you chuse to nourish the child, for that reason to incumber yourself with the contingencies of his future character. * * * * * * * * *

"Yours,

"T. D."

FROM MR. DAY TO MR. EDGEWORTH.

On the same subject.

* * * * "Relative to the principal subject of your letter, I have no more to say, than that I sincerely hope you will never from your own experience change your opinions; at the same time I must candidly confess, that, in spite of the good sense and argument, with which you defend them, I consider them, to a certain degree, as different results in an arithmetical question, which I have examined with so much attention, as scarcely to leave a possibility of being mistaken. Neither do I observe in your letter, that you seem to have entered into the most material parts of mine, the probable inconvenience such a guest might occasion in a large family of females, and the method in which you

would provide for him afterwards, when you had according to his idea given him so many claims upon your generosity. Neither do I see, according to the most favorable result, what great good such an adoption is to produce to yourself. Is not the family you already possess sufficiently large, to employ your cares, your anxieties, your affection, your understanding, and your fortune? Will your children thank you for introducing an alien, who is hereafter to divide with them a portion, that, depend upon it, several of them will think sufficiently scanty? And although every reasonable man, where there is just occasion, should learn to bear the discontents of his own family with fortitude, is there any occasion to add fuel to the flame? or light it where it did not exist before? Do you consider, that in order not to repent of such an action, and only to feel yourself as eligibly situated as you were before, it is perhaps necessary, that, together with good behaviour in yourselves and children, you should find yourself nearly in the same health, spirits, and habit of exertion for perhaps sixteen or twenty years to come; and that the slightest change in any of these circumstances may make you repent your plan, the inconveniencies of which are sure to stick by you, like a caustic upon an ulcer, and perhaps eat down to the very bone? There is also another *little* consideration, which it does not become me to say much about—and that is the dispositions of Mrs. E. She, no doubt, will assent to whatever you propose; but it is not probable, with two families of your children to manage, and an increasing one of her own, she will rejoice much at the prospect of another pupil at Edgeworth-Town.

"I believe no maxim of life to be more incontestably true than this, that all the solid inconveniencies of our conduct stick by us, and increase; while all the fine sentiments, the passions, and the enthusiasm, so entirely evaporate, that we sometimes find a considerable difficulty, when we only endeavour to recollect what could have been our feelings upon some remote occasion.

"Adieu, my dear friend, believe me with
"great sincerity, yours,
"T. D."

FROM MR. EDGEWORTH TO MR. DAY.

"MY DEAR FRIEND,

* * * * * * * * * * "With respect to *your last letter*, as it is your request, the part of it which you mention shall be committed to the flames, if you continue to desire it; but the letter contains so much good sense, and so much kindness, that it shall have a respite till I hear again. All your letters, and those of all my friends, are directed to be returned to them, by my executors, if they wish to destroy them; if not, they are to be given to such of my children as the writers shall think proper.

"The intimate correspondence of real friends, if continued for any length of time, is a treasure of experience for our posterity. The changes which time insensibly produces in the sentiments of the same persons—the different views in which the same objects appear to men of different tempers—the predictions which are so frequently pronounced, and their accomplishment, or fallacy, appearing at once before the mind, afford an interesting lesson to those, who are acquainted with the persons who wrote them.

"Several of my family's letters have come to my hands, and they were not only very entertaining, but very instructive, though they contained neither fine writing nor profound investigation.

"None of my children are ignorant of the esteem, which I have for the gentleman *in his waistcoat* (whose picture is over the sofa in the room where we sit); and his letters, some thirty years hence, will be read with great curiosity, and, in my opinion, to very good purpose, if they are permitted to remain in the family.

"You say, with great truth, my dear friend, that every man's own experience must have taught him such lessons of the unreasonable and encroaching disposition of mankind, as would exceed any description; but we have, likewise, met with gratitude, sincerity, and benevolence, and from every rank of people we have had returns of kindness for the favors that we have conferred; and, if we had never met with any particular instances of regard from individuals whom we had obliged, we have been fully repaid for such of our benevolent exertions, as have been properly directed, by the general esteem of mankind, and that involuntary respect, which is paid to every virtue that is steadily practised with simplicity, and which arises neither from weakness nor affectation, or which cannot be attributed to any selfish motive.

"I do not mean to controvert your argument; I wish only to soften your conclusion: I know, and have ever been forward to assert, that it requires more ability to do good to a single individual, than to torment or ruin twenty; and, perhaps, the philosopher, with the disposal of thousands, would find himself as much circumscribed in his power to do good, as to overcome any other of those obstacles, which Providence has opposed to the exertions of mankind. In a certain mill-horse circle we must grind; and, whenever we attempt to quit the narrow limits of our daily course, we are plucked back by the cord that holds us to the centre, or feel the lash of our invisible driver. As to the subject in question, I feel the force of your remonstrances.

" He, who has attempted to educate so many, must be well acquainted with the difficulties which attend, and the disappointments which follow, the employment.

" If other interests are blended with his own, the difficulties and vexations, which are likely to ensue, should make every man of prudence pause, with deliberation, before he proceeds; but I do not think these difficulties insurmountable, or these vexations intolerable.

" Industry and patience can do much; and the habits of exertion, in which I have passed the greatest part of my life, encourage me to fresh activity. * * * * *

" Yours affectionately,

" R. L. E."

FROM MR. EDGEWORTH TO MR. DAY.

In answer to his complaining of the ingratitude of some inferior person, to whom he had shewn kindness.

* * * * * * You have certainly seen repeatedly, that persons of low habits are never thankful for improvement, comfort, or convenience; but are caught and chained to you by finery, and an introduction into higher life, and greater airs and graces than they had been used to.

" Young people cannot, whilst they are young—and fools cannot, even when old, estimate life with any tolerable accuracy. Their common error is, to attend more to what they hear, than what they feel. And, as they hear almost every body talk with rapture of rank, public diversions, dress, and equipage, they place their hopes of happiness on these objects; and become, at length, so accustomed to pursue them, that, long after they are undeceived as to their real value, they continue the chase, without the slightest appetite for its object.

"We have had, my dear friend, very different fortunes in life. You began with more sanguine hopes of friendship. Not having lived with any person older than yourself, whose abilities you relied upon, you did not hear with the same ears the continual admonitions, that the old dispense upon the frailty of friendship. I have had very few friends; and those I chose amongst my equals in fortune: I remember well, in a lane, near Hare-Hatch, my foretelling, that our attachment was likely to continue, because it was probable, that we should never be in very different stations of life, and, therefore, that we should not be separated by interest or fashion. All this family send their best wishes to you and to Mrs. Day.

"I am, your sincere Friend,

"RICHARD LOVELL EDGEWORTH."

CHAPTER V.

A VERY short time after the preceding letter was sent, my father received one with the usual post-mark of Mr. Day's letters; but the direction was written in a stranger's hand, with "*to be sent immediately*" on the cover.

The change in my father's countenance, when he opened it, I never can forget.

It brought the dreadful intelligence, that Mr. Day was killed by a fall from his horse. This excellent man was, at last, a victim to his own benevolence. Having observed, that horses suffer much in the breaking, from the brutality of common horse-breakers, he had endeavoured, by gentle means, to train a horse for himself; but it was not well broken. It took fright at some one winnowing corn near the road, plunged, and threw him. He had a concussion on the brain; never spoke after his fall; and in less than a quarter of an hour expired!

This account was received from one of Mrs. Day's nephews, who concluded his letter with these words:—

"You, Sir, and Mr. Keir, are the persons whom Mr. Day has selected from his friends, to be his executors, as possessing the greatest probity and superiority of understanding. Mrs. Philips (Mr. Day's mother) and my aunt (Mrs. Day) have commissioned me to write to you, begging you will come over with all possible expedition, after the receipt of this letter; they not being able to settle any of his affairs, till they have the pleasure of seeing you."

To this my father replied.

"Sir, *Oct.* 8, 1789.

"My wife lay in last night, but nothing shall delay me from the melancholy duty, to which I have been appointed by my friend.

"Your obedient servant,

"R. L. Edgeworth."

He set out instantly, but before he sailed for England he had the prudence to enquire at the post-office in Dublin, if there were any English letters for him. There was a second letter from Mrs. Day's nephew, apologising for the precipitancy of his first—hoping my father had not taken a useless journey—saying, that he was not Mr. Day's executor, that the first letter had been written merely from his aunt's having taken it for granted, that her husband had executed a new will, of which he had spoken to her, in which he had appointed Mr. Edgeworth and Mr. Keir his executors; but that, on search being made, no such will was to be found;—that one executed in the year 1780 was the only testament which appeared; that this was lying open in his bureau; and that by this will he left every thing to Mrs. Day, with the exception of a few legacies, appointing her sole executrix. Her nephew added—"Mr. Day has not once mentioned your name, Sir, in the will." To which my father replied:

"Sir,

"I have this moment received your letter. If I did not think I possessed a place in Mr. Day's heart, I should be mortified; but I feel no disappointment in not being mentioned in his will.

"I have an excellent picture of Mr. Day by Wright;

if Mrs. Day should wish for it, it is at her service; she may be sure no other person shall have it, but one of my own children. Give her this message, with my best respects. I have called my new born son Thomas Day, and I hope his name may excite him to imitate the virtues of my excellent friend.

"I am, &c.

"RICHARD LOVELL EDGEWORTH."

It was fortunate, that my father did not sail the day he had intended, for the vessel, in which he was to have taken his passage to Parkgate, was lost.

An amazing defalcation appeared in Mr. Day's property. It proved nearly twenty thousand pounds less, than he had stated it to my father a few years previous to his death. At the time of the American war he had apprehended, that there would have been a national bankruptcy; and under this dread he had sold out of the stocks, and had consulted my father as to the best mechanical means of concealing money. A very considerable sum had been buried under the floor of the study in his mother's house at Barehill. This, to her knowledge, Mr. Day afterwards took up, and placed again in the public funds at the return of peace.

Considering the deficiency that appeared in the property, my father suggested, that some of his money was still concealed, and offered to go over to England, to assist in a search; but the offer was declined; he was assured, that the most diligent search had been made, and that nothing had been found. This is certain, and ought to be stated for the honor of Mr. Day's memory, that, notwithstanding any expressions in his last letter, seemingly tending more to misanthropy than beneficence, he had acted in the most bountiful manner. He had given away by hundreds and thousands, to an amount which one of his intimate friends told Mrs. Day, "that even she, with all her knowledge of his benevolence, could scarcely imagine." Whatever might have been the defalcations in his fortune, and whatever its amount at his death, there was no doubt of his having intended to have made a new disposition of his property.

Mrs. Day wrote to my father, inquiring if he knew of any legacy intended for himself. It had happened, that Mr. Day had before his marriage promised to leave to his friend his library, which was very extensive; but my father would accept from Mrs. Day only some mathematical instruments, which to her were useless, and to him valuable, from old recollections of former times. The library he neither claimed nor would accept, because he thought, that it should now belong to Mrs. Day, to whom, both from her literary tastes, and from her affection for her husband, it must be peculiarly dear. It is worth while to mention this circumstance, in order to give the words, in which Mrs. Day expressed her sense of this attention to her feelings, and her strong and tender affection for her husband.

"MY DEAR MR. EDGEWORTH,

* * * * * * * * "I will ingenuously own, that of all the bequests Mr. Day could have made, the leaving his whole library from me would have mortified me the most, indeed more than if he had disposed of all his other property, and left me only that. My ideas of him are so much associated with his books, that to part with them would be, as it were, breaking some of the last ties, which still connect me with so beloved an object. The being in the midst of books he has been accustomed to read, and which contain his marks and notes, will still give him a sort of *existence* with *me*. Unintelligible as such fond chimeras may appear to many people, I am persuaded they are not so to you. * * * *"

Generous people understand each other. Mrs. Day, of a noble disposition herself, always distinguished in my father the same generosity of disposition. She had, she said, ever considered him as " the most purely disinterested and proudly independent of Mr. Day's friends." She therefore consulted him in all her affairs after she became a widow, and he had the satisfaction of seeing, that she kept every promise which her husband had made, and, as far as she was able, fulfilled even the expectations that he had raised.

She did not forget one individual in particular, who had peculiar claims to her sympathy: one whom Mr. Day had voluntarily taken under his protection from childhood; whom he had educated to be his wife; whom he had loved, and who had always expressed for him the most respectful and grateful attachment—Sabrina Sidney. Those who were interested in the early part of her history must wish to know, what became of Sabrina after Mr. Day's marriage.

She went to reside with a lady in the country, and had lived retired for some years, till she was addressed by a gentleman, who had known her from childhood. The reader may recollect its being mentioned, that, when Mr. Day first selected her from a number of other children, his friend, Mr. John Bicknel, was present; and, in fact, was the person who decided his choice in favor of Sabrina. As she grew up, however, Mr. Bicknel thought so little of her, that he often expressed to my father his surprise, that "his friend Day could be so smitten with Sabrina—he could not, for his part, see any thing extraordinary about the girl, one way or other." When my father praised her voice and manner, Mr. Bicknel only shrugged his shoulders.

Time went on, till Mr. Day's passion for Sabrina, as it has been seen, was sacrificed to philosophy, or was the victim to some trifling misunderstanding. Mr. Bicknel then pitied her, but rejoiced in his friend's more suitable marriage with Miss Milne. Of Sabrina Mr. Bicknel saw nothing more for many years afterwards. He lived in London, partly engaged by pleasure, and partly pursuing his profession of the law. That he was a man of shining talents, his share in the "Dying Negro" is alone sufficient to prove—he had not only poetic taste, but great wit and acuteness. His competitors at the bar, many of whom are now exalted in their profession, were at their first setting out esteemed much inferior to him in abilities. But he had some of the too usual faults of a man of genius; he detested the drudgery of business. He is said to have kept briefs an unconscionable time in his pocket, or on his table, unnoticed. Attorneys complained, but still he consoled himself with wit, literature, and pleasure, till health as well as attorneys began to fail. Then he thought more seriously; and, considering that it would be a comfort to secure a companion for middle life, and a friend, perhaps a nurse, for his declining years, he determined to marry. Suddenly he recollected Sabrina; inquired if she was living, if she was married or unmarried, and asked how she had conducted herself during the long interval since they had met. He found, that she was unmarried, living with a lady in the country, and that she had uniformly behaved so as to deserve the respect, and to conciliate the regard of all who knew her. Mr. Bicknel went down immediately to the part of the country where she resided.—Saw her, and saw her with very different eyes from those, with which he had looked upon her formerly—fell desperately in love—proposed, and was conditionally accepted. *Conditionally*, for Mr. Day, her friend and benefactor, was to be consulted. She declared, that she would not marry any man, without his consent. My father, whom she considered also as a guardian, was consulted: he was surprised, when he first heard of this intended marriage; and when he perceived by the tone of the letters, that Mr. Bicknel was so desperately in love with a lady, whom he had, for so many years, seen with indifference. His office of guardian had long since ceased; the parties were at years of discretion; and, as their friend, he saw no objection to the match, except that Mr. Bicknel's health might fail, and that his application might not be sufficient to secure, by his profession, a maintenance for a family. This objection appeared much more alarming to Mr. Day, than to my father, who always looked to the hopeful side of human affairs; and who believed, that no motive could be stronger or more likely to make a man exert himself, than the desire of providing for a woman he loved.

Mr. Bicknel and Sabrina married; and more than half my father's hopes were realised. Mr. Bicknel did exert himself, and Sabrina made him an excellent wife. In many of Mr. Bicknel's subsequent letters, he spoke of his wife and children, with all the delight of the most happy husband and father. His health, however, failed—he lived scarcely three years after his marriage, and poor Sabrina, after this short period of felicity, was left unprovided for, with two infant sons. Some thought her more unhappy for the felicity she had transiently enjoyed. But this was not my father's doctrine. Two years of happiness he thought a positive good secured, which ought not to be a subject of regret, and should not embitter the remainder of life. Indeed, the system of rejecting present happiness, lest it should, by contrast, increase the sense of future pain, would fatally diminish the sum of human enjoyment; it would bring us to the absurdity of the stoic philosophy, which, as Swift says, "would teach us to cut off our feet, lest we should want shoes."

Mrs. Bicknel never repented of her marriage. Many distinguished characters at the bar, friends of her husband, after his death shewed their regard for him by assistance and kindness to his widow. She exerted herself in a meritorious manner. She maintained independence, and a most respectable character; made for herself many new friends, and preserved all those, who had been early attached to her, and to her husband. My father's attention to her welfare and interests continued constant and energetic, even to the last hours of his life. When these pages were shewn to her, she exclaimed—"You have not said enough—you *cannot* say enough of your father's kindness to me."

Anxious to pay honor due to the memory of his friend, my father, soon after Mr. Day's death, determined to write his life. At that time he was more in the habit of speaking than of writing. It was his custom to throw out, in conversation, his first thoughts upon any subject on which he was intent. I wrote them down, either in my own words, or in his. The beginning of his life of Mr. Day was spoken to me one morning when we were out riding; and the moment I came in, I wrote down the following words, which, I can affirm, were as nearly as possible as they came from my father's lips.

"In the first emotions of grief and affection, which we feel for the loss of a friend, it is natural to imagine, that, could we represent to others the image, which exists in our own mind, we should necessarily command their sympathy in our sorrow for the dead.

"We fondly persuade ourselves, that the private virtues and talents of our lost friend would excite general enthusiasm; that his temper, character, and even his personal peculiarities, must become interesting; and that every circumstance of his life is worthy to be recorded.

"Under the influence of such feelings, many persons are induced to become biographers. But private lives, like family pictures, are valuable only to a few individuals, and, like these, are gradually consigned to oblivion. To preserve a portrait to posterity, it must either be the likeness of some celebrated individual, or it must represent a face, which, independently of peculiar associations, corresponds with the universal ideas of beauty. So the pen of the biographer should portray only those, who by their *public* have interested us in their *private* characters; or who, in a superior degree, have possessed the virtues and mental endowments, which claim the general love and admiration of mankind.

"Immediately after the death of the late Thomas Day, when I first thought of writing his life, perhaps I might have been under the influence of the partial feelings I have described. But the coolest reflections of my understanding have determined me, to pursue the first impulse of my heart.

"Mr. Day was indeed distinguished for the noblest private virtues, but more especially as a philosopher and politician; as a public character he claims the honor of public eulogium.

"Perhaps it will be thought, that, as his intimate friend, I should not trust myself with such a task; but I dare to rely upon my own impartiality; certain that I shall feel proud, to produce the faithful likeness of such a man, without any temptation to correct nature, either from tenderness or presumption."

When my father had nearly completed this life of Mr. Day, hearing that another friend, Mr. Keir, was engaged in a similar undertaking, he gave up his own intentions, and sent to him whatever materials he thought would be of use to him.

TO MR. KEIR.

"My dear Sir, *Edgeworth-Town*, 1790.

"I send you the papers which you desire. It is proper to mention, that I had Mrs. Day's approbation of my undertaking, conveyed in the most obliging manner. But I lay down my pen, fully convinced, that with the same materials you will do much higher honor to your friend's memory.

"The anecdotes which I send you are very few; but they are all I could select to suit your plan, as we differ so materially in our ideas of private biography. You believing, that nothing but what concerns the public should be published; I thinking, that to entertain mankind is no inefficacious method of instructing them. When Mason was reproached by somebody for publishing the private letters of Gray, he answered, 'Would you always have my friends appear in full dress?' I might quote Plutarch as well as Mason in support of my opinion; but I am sure you must perceive, my dear Sir, that I am not willing to enter into any literary competition with you, well knowing my inferiority.

"R. L. E."

Again he says:

TO MR. KEIR.

(Written some months afterwards).

"MY DEAR SIR,

"It continues my very earnest desire, that the papers which I sent to you should become of service in the publication you had undertaken.

"From the moment that I was informed of Mrs. Day's application to you for a life of her husband, I laid aside the thoughts of publishing what I had written. And though she very obligingly pressed me to proceed, intimating, as you had been so kind to do, that our plans would not interfere; I was well convinced, that Mr. Day's fame was in better hands than mine; and that, though my wishes to do justice to his conduct and character could not be surpassed by those of his most enlightened friends, something more than warmth of attachment is requisite in an editor and an author; and I felt sufficiently, that I had little else to rely on.

"The papers I sent are therefore freely at your service; but if either their bulk or their contents are not suited to your plan, I will hereafter endeavour to complete the life which I had begun, and prefix it to a collection of his letters, if Mrs. Day should approve. * * * * *

"Believe me to be, &c.

"R. L. Edgeworth."

At this time, and long after this letter was written, I could not help regretting, that my father's life of his friend, warm from his heart, was suppressed. But all was for the best. I find, that most of the interesting and characteristic circumstances of Mr. Day's life are preserved in my father's own memoirs; and there the characters of the two friends, so different in tastes, yet so agreeing in principle; so opposite in all appearance, yet so attached in reality; mutually illustrate and do honour to each other. Mr. Day's letters in later life further tend to portray his character to the last days of his existence, and they do honor to his private virtues. To his public conduct and to his writings Mr. Keir has done ample justice.

It is remarkable, that Mr. Day's fame with posterity will probably rest solely upon those works, which he considered as most perishable. He valued, in preference to his other writings, certain political tracts; but these, though finely written, full of manly spirit and classic eloquence, have passed away, and are heard of no more. While his " History of Sandford and Merton," and even the story of " Little Jack," are still popular. Wherever children are to be found, these will continue interesting and useful from generation to generation. For the same reason, because true to nature and to general feeling, his poem of " *The Dying Negro*" will last as long as manly and benevolent hearts exist in England.

The health of Mrs. Day had, previous to her husband's death, been precarious; it afterwards rapidly declined. My father, under the apprehension that her constitution was suffering, and might be fatally injured by want of exertion, wrote to her the following letter:—

"MY DEAR MRS. DAY, 1790.

"Your letters are very kind, and I assure you sincerely, that your conduct towards me has given me more pleasure, than any thing has given me since the loss of my friend.

"I am, however, my dear Madam, sensibly concerned at the continuance of your despondence: you will remember, that hitherto I have not attempted to use any argument, nor employed any persuasion, to excite the powers of your mind to exertion. I thought rest and time were necessary; but rest and time may fix habits of sorrow, and weaken every excellent faculty of your soul.

"I have suffered a loss, which was to me as severe as any loss could be to any person of common judgment and affections; and I always perceived, that the idea most prevalent in my mind was, to act and govern myself as she would approve, if she could secretly read my soul, and see my actions. Her dying commands to me all sprang from the generous wish, that I should be happy, though she must cease to be the immediate source of my felicity.

"Ten years are nearly elapsed since I lost her, and her merit, fondness, wisdom, attention to my children, sympathy with all my feelings; the similarity of our tastes, the unbounded confidence which subsisted mutually between us; have never for one day been absent from my grateful remembrance. She breathed her last, smiling, in my arms. Her lifeless form was laid, by her sacred orders, with my own hands, in the last repository of mortality. I attended her to the very grave, and when I saw her covered for ever from my sight, I felt the pangs which nature has made inevitable. But from the moment that I was capable of reflection, I endeavoured by every exertion of my soul, to regain not only tranquillity, but happiness.

"The best of parents was lost to children, that promised to become happy in themselves, and useful to society. The best of wives was lost to a husband, capable of feeling the high value of such a woman. But I knew, that, were my beloved Honora to see me drowned in unavailing sorrow, incapable of the duties of a father and of a man, she would feel indignation at my weakness; and far from accepting the tribute of my grief as a sacrifice due to her exalted merit, she would think herself degraded by lamentations, that are bestowed by common characters upon the commonest occasions.

"Whenever that hour comes, when I must separate from a wife, who far exceeds the promises of her generous sister, and from the intelligent gratitude of my children, I shall conjure them, to shed no tears over my tomb, except such as burst from the first passion of shocked humanity; but to supply my place in society, to carry on the designs, which I had begun with their approbation—to continue that union, which I had formed amongst them for their mutual happiness—to honor me by their actions—to rejoice in remembering me—to speak of me, if I deserve it, with exultation, not with sorrow—to impute my faults to defects in my education: to remember, only to avoid them. To believe from my assertions, perhaps also from their own observation, that I had made, even late in life, improvements in my temper and character; and never to forget, that they might do the same every succeeding year of their lives. This is the honor, which I hope for after death. Let this be the affectionate remembrance of my friends.

* * * * * * * * * * *

No more remains of this letter; nor does

it appear what effect it had upon Mrs. Day's mind. She survived her husband about two years. When my father saw her some time after the date of this letter, though her health was impaired, it was not in such a state, as to create in his mind, or in that of any of her friends, apprehension of danger. But a few weeks after parting with her, we were shocked by hearing of her death.

Of her exemplary conduct as a wife, and as a wife peculiarly suited to Mr. Day, my father has given his testimony strongly; and a single passage in one of her letters, after she became a widow, sufficiently proves the warmth and tenderness of her attachment. So that it cannot be doubted, that her affection stood the trial of all Mr. Day's peculiarities, and in particular of that secluded manner of life, which he thought essential to wisdom, virtue, and happiness.

I cannot quit this subject without repeating what my father has from his own experience asserted, and what was the result of his reflections more decidedly as he advanced in life, that he would never advise any married people, to follow the same plan. Perhaps this caution may seem superfluous in the present state of manners, where people in general are more inclined to be too fond of the dissipation of the world, than prone to retirement; but there are still ingenuous minds of strong affections, and a romantic cast, who will be struck with Mr. Day's noble character, and who might consequently be misled in this particular by his example.

CHAPTER VI.

NOT long after the death of his friend Mr. Day, my father was threatened with another, and a still more severe misfortune, the loss of the only daughter left to him by his beloved wife Honora. From her infancy she had been the object of fond hope. Of high promise, both of personal beauty and intellectual excellence, incessant care had been taken in her education, and of the effect of this it may be useful, and not uninteresting, to the public, to record some particulars. Very early she had shewn a power of abstracting her attention, a capacity for mathematical studies, and facility of accurate definition, uncommon in a child. One instance of this I will mention, because any example, however inadequate, is more satisfactory than general assertion. When she was about seven years old, and had just *heard,* not *learned by rote,* the definitions of a line, a square, and a cube, and had been told what was meant by a body moving through the air, and describing a figure as it moves, she was asked, by her father, the following question:

" If a line move its own length through the air, so as to produce a surface, what figure will it describe?"

She answered,—" *A square.*"

She was then asked,—

" If that square be moved downwards or upwards in the air, the space of the length of one of its own sides, what figure will it, at the end of its motion, have described in the air?"

After a few minutes silence, she answered,—" *A cube.*"

Mrs. Honora E. had been remarkable for strong powers of reasoning, and to some it might appear, as if the daughter inherited her mother's scientific taste, and serious turn of mind. Little Honora did not early shew vivacity, imagination, or invention; and some of my father's friends, who observed this, and who saw her exercised in mathematical reasoning, feared that education, combining with her natural bias, would render her, if I may use the expression, too *reasonable*, or too reasoning. They feared, that whatever power of imagination she might possess would be repressed, or never developed :—those who believe in natural genius were far from expecting that her education, however conducted, could produce excellence of a kind different from that, which nature seemed peculiarly to promise. Here was a fair trial of what could be done.

Abiding by his opinion, that, if the faculty of attention be early cultivated on any one subject, the acquired command of this power may be turned afterwards successfully to whatever object is desired, he pursued his plan steadily. Satisfied with Honora's command of attention on subjects of reasoning and science, he, without anxiety or precipitation, endeavoured to turn her attention to literature; and to that sort of literature, which pleases and excites the imagination. He took delight himself in ingenious fictions, and in good poetry; he knew well how to select what would amuse and interest young people; and he read so well, both prose and poetry, both narrative and drama, as to delight his young audience, and to increase the effect upon their minds of the interest of any story, or the genius of any poet. From the Arabian Tales to Shakspeare, Milton, Homer and the Greek tragedians, all were associated in the minds of his children with the delight of hearing passages from them first read by their father.

He was an enthusiastic admirer of the ancient classics—of Homer and the Greek tragedians in particular. From the best translations of the ancient tragedies he selected for reading aloud the most striking passages, and Pope's Iliad and Odyssey he read several times to his family in certain portions every day. By these means the young Honora's ear and taste for poetry were early formed on the best models, and her love for literature was powerfully excited. The serious and sublime tone of the ancient poets, and of the Greek tragedians, seemed peculiarly to suit her temper and taste. After having heard my father read *Antigone*, she learned by heart one of the scenes*, and recited one of the speeches of Antigone so admirably, as could

* Beginning with—

"Wert thou to proffer, what I do not ask,
"Thy poor assistance, I would scorn it now.
"—Act as thou wilt: I'll bury him myself," &c.

leave no doubt, in the minds of all who heard her, of the powerful effect it must have produced on her feelings and imagination.

Some time afterwards her attention was turned from tragedy to comedy by the acting of that little play, which the reader may recollect is mentioned in one of my father's letters to Dr. Darwin; and in this she succeeded so as to please and amuse her audience, and to fulfil my father's object of diverting her powers into a new course, and exercising her versatility of mind. What success resulted from the general cultivation of her taste for literature, and how far her powers of invention and imagination were developed by the course of education pursued, the public has had before them, in one slight instance, some means of judging, in her first and last literary effort, a little tale (Rivuletta), written when she was about thirteen. Such as it is, it was entirely her own. It was written during my father's absence from home, in the hope of amusing him on his return. Not one word in it was ever altered by him, or by any one.

As Honora grew up, her improvement in every respect was proportioned to this promise of talent. She was indeed a most uncommon and superior creature. Her beauty was such, that it struck all who saw her. Something of serious simplicity and dignity in her manner added to its effect. I have no fear, that any one who ever knew her should charge me with exaggeration in this description. How fondly she was beloved—how highly valued by her parents—may be imagined; and how much of happiness her father must have expected from seeing in her the fruits of her education!—But in the midst of these hopes, before she was fifteen, she was seized by hereditary consumption. From the beginning of her illness, experience of her mother's fate made her father too fully aware of the nature of her disease.

A few months before her death, in one of his letters to Mrs. Day, he says:

* * * "The loss of my best friend must be followed by the loss of my most excellent daughter Honora. Her ripened beauty, her cheerful serene temper, uncommon understanding—all the hopes of her family, by all of whom she is admired and adored—the expectations of all who have ever seen her—must now be blasted. The hand of hereditary disease is upon her, which must soon be inevitably followed by the hand of Death. With the same fortitude which her incomparable mother possessed, she bears the present, and prepares for the future."

Honora died just when she had attained her fifteenth year. It was the first loss of a child my father had ever sustained.

In sorrow the mind turns for comfort to our earliest friends. He went to that sister, whom he mentions in the first part of these memoirs as the favorite companion of his childhood. Their friendship continued a blessing to both in every circumstance of life. With her he had now all that could be done for his consolation by sympathy, by the strong charm of similarity of temper and character, and the stronger charm of association with scenes of youth and early affection. But, as he said, " For real grief, there is no sudden cure. All *human* resource is in time and occupation."

DR. DARWIN TO MR. EDGEWORTH.

" DEAR EDGEWORTH, *Derby*, 1790.

" I MUCH condole with you on your late loss. I know how to feel for your misfortune. The little Tale you sent me is a prodigy, written by so young a person, with such elegance of imagination.

" *Nil admirari* may be a means to escape misery, but not to procure happiness. There is not much to be had in this world—we *expect* too much.

" I have had my loss also! The letter of Sulpitius to Cicero is fine eloquence, but comes not to the heart: it tugs, but does not draw the arrow. Pains and diseases of the mind are only cured by Time. Reason but skins the wound, which is perpetually liable to fester again.

"I am much obliged to you for your second hygrometer*, which is very ingenious. Yet I prefer the former, as being simpler, easier to make, and easier to compare with some third criterion, as by soaking the wood first, and then baking it, and thus making two points of comparison. This also is more certain, than where the friction of pullies is concerned.

"Your three wards were a week with us at Easter. I mention it, to add, that I think them all fine, sensible, good humored boys; and they seemed so grateful for the little notice taken of them by Mrs. Darwin that she quite admires them.

"I will some time make a shelf for your animal to walk upon, and he shall have a walk for a year or two, at two inches a month; but I prefer old *long back* for a race, and would wager him against *brazen-wheels* for a cool hundred.

"I should be very happy to see you this summer, if any thing here can attract such a comet from its orb, and bring any of your family. "Yours truly,

"E. D."

The first, the only agreeable thing, which I recollect about this time, was the arrival of Dr. Darwin's Botanic Garden. My father's delight and enthusiasm, when he read it, his feeling for poetry, and for his friend, can be expressed only in his own words:

* See Note to line 131 of the 3d Canto of the Loves of the Plants.

TO DR. DARWIN.

" *Edgeworth-Town*, 1790.

" It is sometimes, my dear Doctor, a very difficult task for a man to know what to say, when he reads a work written by a friend. His honesty and his affection are often cruelly at variance; and he must either appear a vile flatterer, or an unkind friend.

This is not the difficulty I feel. I am rather in the ridiculous situation of Lightfoot, in the Fairy Tales, who was obliged to tie his legs to restrain his speed.

I have felt such continued, such increasing admiration in reading the " Loves of the Plants," that I dare not express my enthusiasm, lest you should suspect me of that tendency to exaggeration, with which you used to charge me. I may, however, without wounding your delicacy, say, that it has silenced for ever the complaints of poets, who lament that Homer, Milton, Shakspeare, and a few classics, had left nothing new to describe, and that elegant imitation of imitations was all that could be expected in modern poetry. I have seen nobody since it has been published, except my own family; and amongst my domestic critics, who are not readily pleased, I hear nothing but praise and congratulation.

To have my name in a note is, in my opinion, to have it immortal; and, as Mrs. E. says,

" If it's allow'd to poets to divine,
" One half of round eternity is mine."

" I rejoice at all events, my dear Doctor, that I sent you the progressive hygrometer, before I saw your book; as you will impute my sending it somewhat less to vanity, and somewhat more to kindness.

Montgolfier, Lina, and Gossypia, are perfectly new in their ideas. Montgolfier is sublime:—you have not let him strike his head against the stars like Horace; but have made respectful planets recede, to give him place. I would rather have your praise, were I Montgolfier, or Howard, than all the inscriptions or statues of inscribers and academicians.

" My daughter says, that the manner, in which you mention your friends in your poem, shews as much generosity, as your descriptions shew genius.

" But I will stop.——

" I read the description of the Ballet of Medea to my sisters, and to eight or ten of my own family. It seized such hold of my imagination, that my blood thrilled back through my veins, and my hair broke the cementing of the friseur, to gain the attitude of horror. The ghost in Hamlet, by the by, only raised the unconstrained locks of an ill-combed Dane. To force nature through the obstructions of art, is quite another thing.

" My dear Doctor, I will make this fulsome letter a little more palatable by a few verbal criticisms. I know you do not disdain the labor of correcting your poetry:—Maria says, that, even with the help of genii, man can do nothing without some labour; for Aladdin's lamp required to be rubbed quite bright, before the genius obeyed.

" I remark, that the *quantity* of Landau, b. 1, l. 344, is unusual.—B. 1, l. 159, I believe the axle of a weathercock is usually fixed. If I go farther, don't say, *ne sutor*—. L. 165—I don't like Moore of Moore Hall. It is associated with burlesque. I object to l. 290, the vegetable lamb, ending the line. * * * * * * * *

" If you encourage me, I will give you all my ill-natured remarks on the other cantoes.

" Pray, Doctor, flatter me so far as to remember, that the first lines I saw of this poem excited my enthusiasm; and that whilst I sleep, or curse, over many other descriptive poems, I shout applause, when I hear yours.

Qu. Would not the following ideas suit you?—The creation of climate in hot-houses, and ice-houses—The worship of Mercury to the Sun and Juno, in the thermometer and the barometer—Salmoneus and Franklin—Chaos and the Fata Morgana—The *Spectra* of castles—Swinbourne's Travels.—Have you seen Mr. Keir's life of our friend? In Mr. K.'s letters to me, about Mr. Day's Life, he spoke of my mentioning Sabrina as *impossible.* I now find he has changed his opinion. What can the life of a private man consist of, but of private circumstances?

"The *author* appears best in his works.—Adieu.

"R. L. E.

"P. S. I have just read *Papyra,* and would I had more, on which to praise her! You have immortalized one friend, who feels his own insignificance. Pray rescue Mr. Day from oblivion. He named you as one of the three friends, from whom he had met with constant kindness. Forgive me this once, for not knowing how much paper I should want!"

DR. DARWIN IN ANSWER.

Derby, 1790.

"Your last letter is very flattering to me indeed. It was your early approbation, that contributed to encourage me to go on with the poem. The first part is longer than the second, and two of my critical friends think it will be more popular.

"I wish it were possible for me, to shew you the MS. for your criticisms; but I dare not trust it, especially as I have sold it. I am obliged by your criticisms on the Loves of the Plants. I alluded to Moore of Moore Hall, on purpose to give a little air of burlesque, to shelter myself under. For the same purpose the proem was written:—to '*the festoon of ribbon,*' in the preface, I think you helped me.

"*Landau, Chaos, Vegetable lamb,* were all felt to be

objectionable, but not easily mended. I am glad of any criticisms, and if I can *easily* mend them, I do.

" I intend to write no more verses, but to try a medico-philosophical work next, called Zoonomia.

" Why don't you publish something wonderful, you who have so much invention? A century of new machines, with plates; or a decade, would be an agreeable work, and well received, I dare say. You might propose some as practicable, others as tried, &c. &c. I should send you any, which have occurred to me.—I think such a book would be new in its way, and would procure fame.

" Adieu—I shall hope to see you this summer, and am, with best regards to all yours,

" Your affectionate friend and servant,

" E. DARWIN.

" Pray think of a decade of mechanic inventions, with neat drawings, by R. L. Edgeworth, Esq. F. R. S. M. R. I. A. &c. &c. to the end of the alphabet."

I am glad to have it recorded, under Dr. Darwin's own hand, that my father's approbation of the first lines he saw of the Botanic Garden encouraged the author to finish it. With as much sincerity as he gave praise, my father blamed and opposed whatever he thought was faulty in his friend's poem. Dr. Darwin had formed a false theory, that *poetry is painting to the eye;* this led him to confine his attention to the language of description, or to the representation of that, which would produce good effect in picture. To this one mistaken opinion he sacrificed the more lasting, and more extensive fame, which he might have insured by exercising the powers he possessed of rousing the passions, and pleasing the imagination.

When my father found, that it was in vain to combat a favourite false principle, he endeavored to find a subject, which should at once suit his friend's theory, and his genius. He urged him to write a "Cabinet of Gems." The ancient gems would have afforded a subject eminently suited to his descriptive powers; admitting all his elegance and ingenuity of allusion and simile; employing his classical learning, and bringing into play the passions and imagination; with all of ancient history and tragedy, and all of modern poetry, which would have afforded ample range, and rich materials, for creative fancy. The description of Medea, and of some of the labors of Hercules, &c., which he has introduced into his Botanic Garden, shew how admirably he would have succeeded, had he pursued this plan: and I cannot help regretting, that the suggestions of his friend could not prevail upon him, to quit for nobler objects his vegetable loves. He has, however, done all, that was possible for ingenuity and exquisite versification to accomplish, in embellishing a fantastic subject.

My father differed from most critics in one particular. When his advice was not followed, he was not angry; and whenever it was afterward proved, that he was right, and that his friend was wrong, he never triumphed in his own opinion, but made the best of things as they were, instead of shewing how much better they might have been, if his counsel had been taken. It is evident from his letter to Dr. Darwin, that no one could enjoy more sincerely, than he did, the first burst of applause, with which the public received the Botanic Garden:— and when the fashion changed, when Darwin's poetry was cried down, as violently as it had been cried up, my father was its steady and zealous defender.

I have sometimes seen him, when its merits have been questioned, offer to open the Botanic Garden at a venture, a dangerous hazard for any work, and to read aloud from the first page that might occur. His quick eye selected the best lines in the page, and those to which, in reading, he could give contrast and variety. Thus obviating the chief objections to which it is exposed; the sameness of versification and description. As a critic and a judge, he would have admitted these defects; as an advocate and friend, he knew how to conceal them. It was his opinion, that any clouds which have obscured Darwin's genius will pass away, and that it will shine out again, the admiration of posterity. I have heard him predict, that in future times some critic will arise, who shall rediscover the "Botanic Garden," and build his fame upon this discovery.

My father regretted, that his friend did not take his advice about his poetry; and I must express my regret, that my father did not follow Dr. Darwin's counsel about the decade of mechanic inventions. He might in such a publication have inserted a variety of mechanical and agricultural experiments and contrivances, which he had brought to perfection. It was a plan, that would have peculiarly suited his mind, so fertile in invention, so ready in adapting it to practical use, and so habitually conversant in the detail of the small circumstances, which contribute to domestic convenience.

In his own house, or in the houses of his friends, he was continually devising new means of adding to their comfort. Even in that country, where the idea of comfort, as a foreigner observes, forms a distinguishing part of the national character, few understood this subject better than he did. He had executed, and put to the proof of long and constant use, a variety of small inventions, which, separately considered, are scarcely worth mentioning; but which, altogether, add essentially to domestic order, and every-day enjoyment. In these things, provided the principle and the mechanics were good, he cared little for any appearance, that was to catch the popular eye. After having made and remade innumerable models, sometimes with his own hands, of the same contrivance, when he had satisfied himself that he could improve it no further, he would leave it executed in the roughest form. Even a favorite invention, that clock which he mentions having made when he was a young man, as having tried as a time-keeper at Oxford, and which to my knowledge was going for upwards of thirty years in his own house—he, in spite of reiterated reproaches, neglected to describe for the Royal Society.

Of course it often happened, that the same things, which he had invented, were invented and published by others. Instead of feeling pain, he, however, felt pleasure at seeing the approbation excited by the publicity of those ideas, which he knew had previously passed through his own mind. I have heard him exclaim with delight, on reading the public papers, or reviews, when any circumstance of this kind occurred. His self-complacency was peculiarly gratified in thus being assured of the measure of his own capacity, compared with the inventive ability of the first people of his day. In short, he was careless about fame, to a degree that would hardly be believed by those, who are jealous of every petty rivalship of invention, and who raise the cry of plagiary at the appearance of every resemblance or coincidence of ideas. Even while we were provoked by his losing the just praise due to his inventions, we loved and respected him the more for his magnanimity.

Many circumstances, peculiar to the country and the times in which he lived, interrupted his scientific pursuits, and forced him to attend to less agreeable, but more immediately necessary subjects. About this period, he was called upon to take the management of the embarrassed affairs of a relation. Much of his time was occupied in making out accounts, and settling between debtors and creditors.

In endeavouring to arrange with the creditors, he had of course some difficulties, and was ultimately at considerable loss; but when he attempted to collect what was due of arrears of rent on his relation's estate, the matter became not only difficult, but perilous; for it was his fate, to have to deal with persons calling themselves *gentlemen tenants*—the worst tenants in the world—*middle-men*, who relet the lands, and live upon the produce, not only in idleness, but in insolent idleness.

This kind of half gentry, or mock gentry, seemed to consider it as the most indisputable privilege of a gentleman, not to pay his debts. They were ever ready to meet civil law with military brag of war. Whenever a swaggering debtor of this species was pressed for payment, he began by protesting, or *confessing*, that "he considered himself used in an ungentlemanlike manner;" and ended by offering to give, instead of the value of his bond or promise, "the *satisfaction* of a gentleman, at any hour or place." Thus they put their promptitude, to hazard their worthless lives, in place of all merit, especially of that virtue, by them most despised, perhaps because by them least known—erroneously called *common* honesty. It certainly was not easy to do business with those, whose best resource was to settle accounts by wager of battle with the representative of their deceased creditor; nor was it easy, while inferior persons felt it their interest and ambition to provoke their antagonist, to keep out of discreditable quarrels, by which nothing could be gained, and every thing might be lost. It required not only prudence and temper, but established character, with some weight of family connexions, and the united voice of good friends, to bear him out, at this time, in the cause of justice, when it was on the creditor side of the question.

My father has often since rejoiced in the recollection of his steadiness at this period of his life. As far as the example of an individual could go, it was of service in his neighbourhood. It shewed, that such lawless proceedings, as he had opposed, *could* be effectually resisted; and it discountenanced that braggadocio style of doing business, which was once in Ireland too much in fashion. Such would no longer be tolerated in this part of the country; but such has been: and persons of the sort I have described flourished some thirty years ago, and were among a certain set popular as *men of undeniable spirit.*

In the year 1792, the health of one of his sons obliged my father to leave Ireland, and to go to Clifton, where he and his family resided nearly two years. This was the first time I had ever been with him in what is called *the world;* where he was not only a useful, but a most entertaining guide and companion. His observations upon characters, as they revealed themselves by slight circumstances, were amusing and just. He was a good judge of manners, and of all that related to appearance, both in men and women. Believing, as he did, that young people, from sympathy, imitate or catch involuntarily the habits and tone of the company they keep, he thought it of essential consequence, that on their entrance into the world they should see the best models. "No company or good company," was his maxim. By *good,* he did not mean *fine.* Airs and conceit he despised, as much as he disliked vulgarity. Affectation was under awe before him, from an instinctive perception of his powers of ridicule. He could not endure, in favor of any pretensions of birth, fortune, or fashion, the stupidity of a formal circle, or the inanity of commonplace conversation. His impatience might, at the moment, be properly concealed; yet the force with which it burst out, when the pressure was taken off, gave the measure of the constraint he had endured.

He could not bear the system of visiting, merely to increase the visiting list, or to strengthen the league defensive and offensive of persons, who are to bow and curtsy exclusively to each other in public places. He used to say, that the misery of half those, who are wretched in the higher classes of society, arises from their inability to live out of crowds and public places; and he spoke with much indignation and eloquence against that gregarious spirit, which tends to stupify and enslave individuals, and to debase the national character.

Sometimes, perhaps, he went too far, and at this period of his life was too fastidious in his choice of society; or when he did go into mixed company, if he happened to be suddenly struck with any extravagance or meanness of fashion, he would inveigh against these with such vehemence, as gave a false idea of his disposition. In short, though no man could more enjoy good company, or, I may say, was better calculated to please and shine in it, from his manners, and the life, variety, and originality of his conversation; yet, when the fit of indignation was on him, it might have been supposed, that he was possessed by the spirit of Mr. Day. From long intercourse and habitual admiration of his friend's noble character, he had caught some of that spirit, even while he saw its errors, though no one could feel these more, or mark them better. Once, in speaking of Mr. Day's wasting the weight and power of philosophy, by employing it against trifles, he said, that "the strength of Hercules could not throw a feather further than it can be thrown by an infant."

It is remarkable, that Mr. Day's declamations against fashionable follies had always been borne with patience by his acquaintance; for they were considered as suitable to a grave, professed philosopher, who never looked, dressed, or comported himself like other people. His auditors secretly consoled themselves for his contempt by the belief, that he knew nothing of the world. On the contrary, with regard to my father, they were provoked to find, that one, who could please in any company, should disdain theirs; and that he, who seemed made for society, should prefer living shut up with his own friends and family. An inconvenience arose from this, which is of more consequence than the mere loss of popularity, that he was not always known or understood by those, who were really worthy of his acquaintance and regard. With that truth and candor, which I know he would have approved, I have here mentioned what appeared to me an error of system and practice during some years of his life. In justice to him I must mention,—that Mrs. Edgeworth's taste, as well as his own, was for retirement, and that her health did not permit her, to mix much with the world.

The mode of life at a water-drinking place was not suited to him; and he did not like these two years' residence at Clifton, where he was deprived of the means of employment and interest, which he enjoyed at home, in agriculture and mechanics, and the various business of active life. His mind, however, was not idle. The following letter, from his excellent friend Mr. Wedgwood, may give an idea of the objects, with which he was continually occupied.

MR. WEDGWOOD TO MR. EDGEWORTH.

"DEAR SIR, "*Etruria, Oct.* 24, 1792.

"I AM truly sorry for the occasion of your visit to Bristol Wells * * * * * * * * * * * * . Matlock waters are likewise much celebrated, and you may know, that our good friend Dr. Darwin thinks, that they possess much the same properties.

"I should think, that the properties of the Clifton waters might be heightened by the execution of your very ingenious thought of filling them with a quantity of fixed air, sufficient to make them rival in that respect the waters of Pyrmont. But I believe, that, if it came to be known, it would ruin, rather than increase the exportation trade of this water, with the bulk of consumers. I should like it much, so medicated, for my own drinking, having received great benefit from aërated water; but I have nevertheless my doubts of the plan being generally approved, not because the water would not be mended by the addition, but merely from the idea, that these things ought to be taken *as God sends them,* without our presuming to attempt an amendment of what we do not understand; but you will be so good as consider me now speaking merely as a water merchant.

" I admire your unwearied and prolific genius, ever laboring for the advantage of mankind, and I answer your observations and queries without delay.

" As you observe, the present coverings to our habitations are liable to many objections, and well he deserves the civic crown from his countrymen, who is happy enough to introduce any important improvement, either in their materials or construction. You wish to be informed, whether tiles may not be made with safety lighter than slates, and of a color to resemble them. Perhaps they could not with safety be made lighter than some of our slates now are; they would become too tender, and every little crack and flaw, to which burnt clay will be liable, would go through the tile, and let the water in. Some slates are light even to a fault, and the roofs are often stripped by a blast of wind. Such are the Welsh blue slates. Tiles could not, as you suggest, be easily made to resemble the color of the beautiful Westmoreland slate, by any means I am at present acquainted with: the use of cobalt and a white clay seem necessary. The cobalt would not give its color to any other but a white clay; this coloring material being sold at two guineas a pound, seems to prohibit its use for this purpose. Nor do I know any pleasing color, that could be given to tiles; the red of pantiles is too fiery; our dark colored tiles, falsely called blue, are too dismal: in short, I do not know any thing, that would at all bear looking upon, for color, in comparison with the Westmoreland slate.

" With respect to drawing the tiles out under a cylinder, upon boards, as you propose, and, when sufficiently dry, planing them to a proper thinness, this might no doubt be done; but the workmen could form them quite thin enough, and much more expeditiously, in moulds: besides, after they were stiff enough to bear planing, it seems to me, that the turning of their edges would then be much more difficult, if not impossible. Stourbridge clay is certainly an excellent material for making the backs of grates, or any large pieces, which are to be exposed to a considerable degree of heat. I am glad to hear they are applying it to any of these salutary purposes, and I wish you success in your comparison of it with heated iron.

" Shall I trouble you to present, in a proper moment, my best thanks to Mr. Lovell Edgeworth, for his elegant poetical description of the Barbarini, now the Portland vase. It is highly pleasing to me, to see so large a portion of the father's genius descending to his son; and I hope and trust, that ample time will be allowed for the growth and cultivation of so promising a scion. I need not tell you the pleasure I have received from the perusal of this beautiful description of an unique work of ancient art. You acknowledge it has faults, and I will freely tell you that which strikes me most—the compliment which he introduces to me. It is too flattering an instance of his kind partiality to me and my works.

" Adieu. Believe me with the sincerest regard,

" Your faithful

" J. Wedgwood."

The following letter, addressed to the Roman catholics of the county of Longford, states my father's political opinions on two important subjects; and proves, that, even when absent from Ireland, he was always intent upon what he believed to be for her welfare.

" GENTLEMEN, " *Clifton, Nov.* 1, 1792.

" I TAKE the opportunity of addressing you, as my absence from Ireland prevents me from knowing any better mode of communicating my sentiments to the Roman catholics of the county of Longford.

" Ever since I have taken any part in the politics of Ireland, I have uniformly thought, that there should be no civil distinctions between its inhabitants upon account of their religious opinions. I concurred with a great character at the national convention, in endeavouring to persuade our Roman catholic brethren, to take a decided part in favor of a parliamentary reform. They declined it; and it then became absurd and dangerous for individuals, to demand rights in the name of a class of citizens, who would not avow their claim to them. As soon as you have a probability of success, many pretended friends will clamour loudly in your favour. I wish, before this crisis happens, to declare myself in favor of a full participation of rights amongst every denomination of men in Ireland; and if I can, by my personal interference at any public meeting of our county, serve your cause, I shall think it my duty to attend.

" I am, Gentlemen,

" Your steady friend,

" RICHARD LOVELL EDGEWORTH."

In various visits to London during the two years which he spent in England, my father renewed his acquaintance with literary and scientific friends; and he often met with persons and circumstances, which recalled previous passages of his life. Of these occurrences, which he always made entertaining to us in the narration, I recollect and will mention one; for though it is trivial, it is characteristic. In his youth he had been remarked as a good dancer; and, as he says in his memoirs, he had learned the art from the famous Aldridge. In 1793, nearly thirty years having elapsed since that time, happening to be in a coffee-house in London, he observed a gentleman eyeing him with much attention, who at last exclaimed—"It is he!—Certainly, Sir, you are Mr. Edgeworth?"

"I am, Sir."

"Gentlemen," said the stranger, with much importance, addressing himself to several people who were near him—"Here is "the best dancer in England, and a man to "whom I am under infinite obligations; for "I owe to him the foundation of my fortune. "Mr. Edgeworth and I were scholars of the "famous Aldridge; and once when we prac- "tised together, Mr. E. excelled me so much, "that I sat down upon the ground, and burst "out a crying; he could actually complete "an entrechat of ten distinct beats, which "I could not accomplish! However, I was "well consoled by him, for he invented, for

" Aldridge's benefit, '*The tambourine dance,*' " which had uncommon success. The dresses " were Chinese. Twelve assistants held small " drums, furnished with bells, these were " struck in the air by the dancer's feet, when " held as high as their arms could reach. This " Aldridge performed, and *improved* upon by " stretching his legs asunder so as to strike " two drums at the same time. Those not " being the days of elegant dancing, I after- " wards," continued the stranger, " exhi- " bited, at Paris, the tambourine dance, to " so much advantage, that I made fifteen " hundred pounds by it."

The person, who made this singular address and eulogium, was the celebrated dancer, Mr. Slingsby. His testimony proves, that my father did not overrate his powers as a dancer; but it was not to boast of a frivolous excellence, that he told this anecdote to his children; it was to express his satisfaction, at having, after the first effervescence of boyish spirits had subsided, cultivated his understanding, turned his inventive powers to useful objects, and chosen as the companions of his maturer years men of the first order of intellect.

His friends and companions in invention

now were Watt, Darwin, Keir, Wedgwood: all these he of course visited while he was in England. Besides, he formed new acquaintance with many, who were distinguished for ingenuity and science. At this time I may date the commencement of his friendship with the ingenious, indefatigable, and benevolent Mr. William Strutt, of Derby, with whom he was made acquainted by Dr. Darwin; the common friend of genius and of goodness, which he had the happy talent of discovering, attracting, and attaching.

But while my father was thus refreshing old intimacies, and enjoying new ones, in England, he lost, in Ireland, the man whom, next to Mr. Day, he had always considered as his best friend—the late Lord Longford. The news of Lord Longford's illness had never reached him, till it was accidentally mentioned before him, by a gentleman, who was reading a letter from Ireland. Before my father could receive an answer to a letter which he instantly wrote, offering to return to see his friend, if there was any truth in this report, Lord Longford was no more. His services in the British Navy, and his character as an Irish senator, have been fully appreciated by the public. His value in private life, and as a friend, can be justly estimated only by those, who have seen and felt how strongly his example and opinions have, for a long course of years, continued to influence his family, and all who had the honor of his friendship. The permanence of this influence after death is a stronger proof of the sincerity of the esteem and admiration felt for the character of the individual, than any which can be given during his lifetime. I can bear witness, that in one instance it never ceased to operate. I know, that on every important occasion of my father's life, where he was called upon to judge or act, long after Lord Longford was no more, his example and opinions seemed constantly present to him; he delighted in the recollection of instances of his friend's sound judgment, honor, and generosity; these he applied in his own conduct, and held up to the emulation of his children.

While we resided at Clifton, my father became acquainted with the celebrated Dr. Beddoes; and it is remarkable, that this acquaintance was in consequence of the Doctor's great admiration for the character of Mr. Day. This had induced Dr. Beddoes to seek the acquaintance of Mrs. Day, and of her friend, Mr. Keir. When Dr. Beddoes came to Clifton, with the view of settling as a physician, Mr. Keir gave him a letter of introduction to my father, who was, I believe, his first acquaintance there. My father admired his abilities, was eager to cultivate his society; and this intimacy continuing some months, he had opportunities of assisting in establishing the Doctor at Clifton.

In the autumn of 1793, we heard, that disturbances were beginning to break out in Ireland, and my father thought it his duty, to return there immediately. Our preparations for leaving Clifton seemed particularly to grieve and alarm Dr. Beddoes. During the summer's acquaintance with our family, he had become strongly attached to one of my sisters—Anna. In consequence of the declaration of his passion, and to give her opportunity to see more of him, my father remained some time longer at Clifton. She decided to return with us to Ireland that autumn, to take further time to judge of the permanence of Dr. Beddoes's feelings, and of her own. He had permission to follow her in the spring; and they were married at Edgeworth-Town, on the 17th of April, 1794.

CHAPTER VII.

THE disturbances in Ireland, of which we had seen exaggerated accounts in the English newspapers, did not, at our first return home, appear formidable. Though we were occasionally alarmed by reports of outrages committed by *heart-of-oak boys* and *defenders* in distant counties, and though in our own there were some nightly marauders, yet, upon the whole, our neighbourhood continued tolerably quiet. Towards the end of the year 1794, a rumour of a French invasion spread through the country, raising the hopes of the disaffected, and creating terror in all, who had any thing to lose. The variety of contradictory reports, the difficulty which the higher classes found in ascertaining the truth, increased the embarrassment and the danger. Means of quickly receiving and communicating information were essential. At this time, my father proposed to the Irish government, a telegraph of his invention, and persevered in his efforts, to have it established at his own expense, or in any way government might approve.

The following letters to his friend Dr. Darwin give some view of the state of the country, of my father's intentions with respect to the telegraph, and of his public and private pursuits, during this period.

FROM MR. EDGEWORTH TO DR. DARWIN.

"MY DEAR DOCTOR, *Edgeworth-Town, Sept. 7,* 1794.

"Being, like Sancho, a retailer of proverbs, I remark, 'that it never rains, but it pours.' Your obliging letter came the day after Zoonomia, which was retarded by winds and waves.

"Had it come sooner to my hands, you may be sure, that you would have received my acknowledgements for your present.

"Some of my great, great grand-children will, with laudable vanity, shew the *ex dono* to their envying contemporaries, and exult in the friendship of their ancestor.

"Just recovering from the alarm occasioned by a sudden irruption of defenders into this neighbourhood, *and* from the business of a county meeting, *and* the glory of commanding a squadron of horse, *and* from the exertion requisite to treat with proper indifference an anonymous letter sent by persons, who have sworn to assassinate me, I received the peaceful philosophy of Zoonomia; and though it has been in my hands not many minutes, I found much to delight and instruct me. * * * *

"We were lately in a sad state here—the sans culottes (literally so) took a very effectual way of obtaining power; they robbed of arms all the houses in the country; thus arming themselves, and disarming their opponents. By *waking* the bodies of their friends, the human corpse not only becomes familiar to the sans culottes of Ireland, but is associated with pleasure, in their minds, by the festivity of these nocturnal orgies. An insurrection of such people, who have been much oppressed, must be infinitely more horrid, than any thing that has happened in France; for no hired executioners need be sought from the prisons, or the galleys. And yet the people here are altogether better than in England. The higher classes are far worse*; the middling classes far inferior to yours, very far indeed; but the peasants, though cruel, are generally docile, and of the strongest powers, both of body and mind.

" A good government may make this a great country, because the raw material is good and simple. In England, to make a carte-blanche, fit to receive a proper impression, you must grind down all the old rags, to purify them.

" DOMESTIC OCCURRENCES.—Lovell is in excellent health, and does not fear the Equinox; he goes to Edinburgh next month: he has been employed in building, and other active pursuits, which seldom fall to the share of young men, but which seem as agreeable to him as the occupations of a mail-coachman, a groom, or a stable-boy, are to some youths. I am every day more convinced of the advantages of good education. One of my younger boys is what is called a genius—that is to say, he has vivacity, attention, and good organs. I do not think one tear per month is shed in the house, nor the voice of reproof heard, nor the hand of restraint felt. To educate a second race costs no trouble. *Ce n'est que le premier pas, qui coute.*

" Adieu, my excellent friend: your generous conduct to Dr. Beddoes does not escape me. Give my best services to Mrs. Darwin, whom I admire as much as a man ought to admire the wife of his friend.

" Yours,

" R. L. EDGEWORTH."

P. S. " What you say of ennui in Zoonomia is an admirable observation. The fondness for *life itself* is certainly pro-

* The reader will observe, that this was written thirty years ago; what is here said refers, *I know*, to the political, not to the moral character of the higher classes in Ireland. M. E.

duced by the pleasurable sensations of different movements in our bodies, that have no names; the sum of which counterbalance a great share of pain. Atticus, by starving himself, had destroyed these pleasures so far as to make him unwilling to be at the trouble of living, even after his disease had left him."

FROM THE SAME TO THE SAME.

"MY DEAR DOCTOR, *Edgeworth-Town, Dec.* 11, 1794.

"The Swiss 'mal du pays' is a periodical disease, which comes on after certain intervals. I believe that every body, who has any heart, after a certain time longs to see, and, if they cannot visit, to write to their friends *beyond sea.*

"I have been employed for two months in experiments upon a telegraph of my own invention. I tried it partially twenty-six years ago. It differs from the French in distinctness and expedition, as the intelligence is not conveyed alphabetically.

"I propose to government, to raise a corps of Vedettes, and to station them in fifty or sixty posts, from which a constant correspondence may be kept up with the capital. I also propose to carry my visual communication across the channel to London.

"In your country I hope, that ministers have satisfied themselves, that there was no design to introduce *sans culottism;* and that the discontent, which reasonable men expressed against the flagrant abuses, and insane wars, was a feeling very different from a wish to call old Chaos from the bed of night.

"Here things are very different. The lowest order of the people has been long oppressed; they are ignorant; they are *vrais sans culottes,* and, without prevention, the most horrid calamities may ensue.

"I intended to detail my telegraphs (in the plural), but I find that I have not room at present. If you think it worth while, you shall have the whole scheme before you, which I know you will improve for me. Suffice it, that, by day, at eighteen, or twenty miles distance, I shew, by four pointers, isosceles triangles, twenty feet high, on four imaginary circles, eight imaginary points, which correspond with the figures

0, 1, 2, 3, 4, 5, 6, 7.

So that seven thousand different combinations are formed, of four figures each, which refer to a dictionary of words; by an additional contrivance, seven different vocabularies are referred to—of lists of the Navy, Army, Militia, Lords, Commons, geographical and technical terms, &c., besides an alphabet. So that every thing one wishes may be transmitted with expedition.

" By night, white lights are used—Query, the most economical? I wrote to another friend, but he is as laconic and obscure as Delphos.

" Adieu, my dear Doctor. Lovell braves the winter, and becomes, I hope, every year a life better worth your care.

" Anna is happy with Dr. Beddoes, and the rest of my family flourish: so I hope do all yours, and at their head their excellent and amiable mother.

" Yours,

" R. L. EDGEWORTH."

" I hope Erasmus is growing rich and powerful; as he is not married, what but riches and power can interest him?"

FROM DR. DARWIN TO MR. EDGEWORTH.

" DEAR SIR, *Derby, March* 15, 1795.

" I beg your pardon, for not immediately answering your last favor, which was owing to the great influence the evil demon has at present in all affairs on this earth. That is, I lost your letter, and have in vain looked over some scores of papers, and cannot find it. Secondly, having lost your letter, daily hoped to find it again—without success.

" The telegraph you described, I dare say, would answer the purpose. It would be like a giant wielding his long arms, and talking with his fingers: and those long arms might be covered with lamps in the night. You should place four or six such gigantic figures in a line, so that they should spell a whole word at once; and other such figures within sight of each other, all round the coast of Ireland; and thus fortify yourselves, instead of Friar Bacon's wall of brass round England—with the brazen head, which spoke—'Time is! Time was! Time is past!'

"The death of Mr. Wedgwood grieves me very much; he is a public, as well as a private loss. We all grow old but you! When I think of dying, it is always without pain or fear.

"I think you or Lovell write Latin verses sometimes. I see Mr. Seward (as I believe), has put my head in the European Magazine for the last month, with an account of my life and opinions. I have written the following, to put at the end of the next volume of Zoonomia, which is now printing. Correct the lines, or improve, or amplify.

Currus Triumphalis Medicinæ.

Currus it Hygeiæ,—Medicus movet arma triumphans,
  Undique victa fugit lurida turma mali.—
Laurea dum Phœbi viridis tua tempora cingit,
  Nec mortale sonans Fama coronat opus;
Post equitat trepidans, repetitque Senectus in aurem,
  Voce canens stridulâ, "Sis memor ipse mori!"—

"I kept this letter three or four days, to send you a legal receipt, and now I don't know how to send you one. Mr. Burns, the Quaker, once let me a lease, which began with—'Friend Darwin, honest men need no lawyer, I hereby let you a lease,' &c. So I send you my receipt, as follows * * * * * without technical form.

"Adieu; your sincere and affectionate friend,

"E. DARWIN."

In August, 1794, my father first made a trial of his telegraph, between Pakenham

Hall and Edgeworth-Town, a distance of twelve miles; he found it to succeed beyond his expectations: and in November following, he made another trial of it at Collon, at Mr. Foster's, in the county of Louth. The telegraphs were on two hills, at fifteen miles distance from each other. A communication of intelligence was made, and an answer received, in the space of five minutes. Mr. Foster, my father's friend, and the friend of every thing useful to Ireland, was well convinced of the advantage and security this country would derive from a system of quick and certain communication; and, being satisfied of the sufficiency of this telegraph, advised, that a memorial on the subject should be drawn up for government. Accordingly, under his auspices, a memorial was presented, in 1795, to Lord Camden, then Lord Lieutenant. His Excellency glanced his eye over the paper, and said, that he did not think such an establishment necessary, but desired to reserve the matter for further consideration. My father waited in Dublin for some time. The suspense and doubt, in which courtiers are obliged to live, is very different from that state of philosophical doubt, which the wise recommend, and to which they are willing to submit. My father's patience was soon exhausted. The county in which he resided was then in a disturbed state, and he was eager to return to his family, who required his protection. Besides, to state things exactly as they were, his was not the sort of temper suited to attendance upon the great, as will appear from the following note.

TO MRS. EDGEWORTH.

"MY DEAREST LOVE, *Dublin, 7th Jan.* 1795.

"My business is taken *ad referendum*, in the Dutch manner. But it was expected, I suppose, that I should pay my court: no court on earth shall detain me another day from those I love, and whom I shall protect as far as my powers of mind and body will permit. The news I hear from the county of Longford makes even my philosophy impatient to return to you. My best love to dear Charlotte Sneyd. My second best in order due. Send the chaise to meet me on Friday.

"R. L. E."

The disturbances in the county of Longford were quieted for a time by the military; but again, in the autumn of the ensuing year, (Sept. 1796), rumours of an invasion prevailed, and spread with redoubled force through Ireland, disturbing commerce, and alarming all ranks of well disposed subjects. My father wrote to Lord Carhampton, then commander in chief, and to Mr. Pelham, (now Lord Chichester), who was then secretary in Ireland, offering his services. The secretary requested Mr. Edgeworth would furnish him with a memorial. Aware of the natural antipathy, that public men feel to the sight of long memorials, this was made short enough, to give it a chance of being read.

Presented, Oct. 6th, 1796.

"Mr. Edgeworth will undertake to convey intelligence from Dublin to Cork, and back to Dublin, by means of fourteen or fifteen different stations, at the rate of one hundred pounds per annum for each station, as long as government shall think proper: and from Dublin to any other place, at the same rate, in proportion to the distance; provided, that, when government chuses to discontinue the business, they shall pay one year's contract over and above the current expense, as some compensation for the prime cost of the apparatus, and the trouble of the first establishment."

In a letter of a single page accompanying this memorial, it was stated, that to establish a telegraphic corps of men sufficient to convey intelligence to every part of the kingdom where it should be necessary, stations tenable against a mob, and against musketry, might be effected, for the sum of *six or seven thousand pounds*: it was further observed, that, of course, there must be a considerable difference between a partial and a general plan of telegraphic communication; that Mr. Edgeworth was perfectly willing to pursue either, or to adopt, without reserve, any better plan, that government should approve.—

Thanks were returned, and approbation expressed.

Nothing now appeared in suspense except the *mode* of the establishment, whether it should be civil or military. Meantime Mr. Pelham spoke of the Duke of York's wish, to have a reconnoitring telegraph, and observed, that Mr. Edgeworth's would be exactly what his Royal Highness wanted. Mr. Edgeworth in a few days constructed a portable telegraph, and offered it to Mr. Pelham. He accepted it, and at his request my brother Lovell carried it to England, and presented it to the Duke from Mr. Pelham.

During the interval of my brother's absence in England, my father had no doubt, that arrangements were making for a telegraphic establishment in Ireland. But the next time he went to the Castle, he saw signs of a change in the Secretary's countenance, who seemed much hurried—promised he would write—wrote, and conveyed, in diplomatic form, a final refusal. Mr. Pelham, indeed, endeavored to make it as civil as he could, concluding his letter with these words:

" The utility of a Telegraph may hereafter be considered greater, but I trust, that at all events, those talents, which have been directed to this pursuit, will be turned to some other object; and that the public will have the benefit of that extraordinary activity and zeal, which I have witnessed on this occasion, in some other institution, which, I am sure, that the ingenuity of the author will not require much time to suggest.

"I have the honor to be,

Dublin Castle, "With great respect, &c.

17*th Nov.* 1796. "T. PELHAM."

Of his offer to establish a communication from the coast of Cork to Dublin, at *his own expense*, no notice was taken. "He had, as was known to government, expended 500*l.* of his own money; as much more would have erected a temporary establishment for a year to Cork. Thus the utility of this invention might have been tried, and the most prudential government upon earth could not have accused itself of extravagance in being partner with a private gentleman, in an experiment, which had, with inferior apparatus, and at four times the expense, been tried in France and England, and approved." The most favorable supposition, by which we can account for the conduct of the Irish government in this business, is, that a superior influence in England forbade them to proceed. "It must," said my father, "be mortifying to a viceroy, who comes over to Ireland with enlarged views, and benevolent intentions, to discover, when he attempts to act for himself, that he is peremptorily checked; that a circle is chalked round him, beyond which he cannot move."

No personal feelings of pique or disgust prevented my father from renewing his efforts to be of service to his country. Two months after the rejection of his telegraph, on Friday the 30th of December, 1796, the French were on the Irish coasts. Of this he received intelligence late at night: immediately he sent a servant express to the Secretary, with a letter, offering to erect telegraphs, which he had in Dublin, on any line that government should direct, and proposing to bring his own men with him; or to join the army with his portable telegraphs, to reconnoitre. His servant was sent back with a note from the Secretary, containing compliments, and the promise of a speedy answer: no further answer ever reached him. Upon this emergency he could, with the assistance of his friends, have established an immediate communication between Dublin and the coast, which would not have cost the country one shilling. My father shewed no mortification at the neglect with which he was treated, but acknowledged, that he felt much "concern in losing an opportunity of saving an enormous expense to the public, and of alleviating the anxiety and distress of thousands." A telegraph was most earnestly wished for at this time by the best informed people in Ireland, as well as by those, whose perceptions had suddenly quickened at the view of immediate danger. Great distress, bankruptcies, and ruin to many families, were the consequences of this attempted invasion. The troops were harassed with contradictory orders and forced marches, for want of intelligence, and from that indecision, which must always be the consequence of insufficient information. Many days were spent in terror, and in fruitless wishes for an English fleet. One fact may mark the hurry and confusion of the time; the cannon and the ball sent to Bantry Bay were of different calibre. At last Ireland was providentially saved by the change of the wind, which prevented the enemy from effecting a landing on her coast.

That the public will feel little interest in the danger of an invasion of Ireland, which might have happened in the last century; that it can be of little consequence to the public, to hear how, or why, twenty years ago, this or that man's telegraph was not established, I am aware:—and I am sensible that few will care how cheaply it might have been obtained, or will be greatly interested in hearing of generous offers, which were not accepted, and patriotic exertions, which were not permitted to be of any national utility. I know, that, as a biographer, I am expected to put private feelings out of the question; and this duty, as far as human nature will permit, I hope I have performed.

The facts are stated from my own knowledge, and from a more detailed account, in his own "*Letter to Lord Charlemont on the Telegraph*," a political pamphlet, uncommon at least for its temperate and good humored tone.

Though all his exertions to establish a telegraph in Ireland were at this time unsuccessful, yet he persevered in the belief, that, in future, modes of telegraphic communication would be generally adopted; and instead of his hopes being depressed, they were raised and expanded by new consideration of the subject in a scientific light. In the sixth volume of the Transactions of the Royal Irish Academy, he published an "Essay on the art of conveying swift and secret intelligence;" in which he gives a comprehensive view of the uses, to which the system may be applied, and a description, with plates, of his own machinery. Accounts of his apparatus, and specimens of his vocabulary, have been copied into various popular publications, therefore it is sufficient here to refer to them. The peculiar advantages of his machinery consist, in the first place, in being as free from friction as possible; consequently in its being easily moved, and not easily destroyed by use. In the next place, on its being simple, consequently easy to make and to repair. The superior advantage of his vocabulary arises from its being undecipherable. This depends on his employing the numerical figures instead of the alphabet. With a power of almost infinite change, and consequently with defiance of detection, he applies the combination of numerical figures to the words of any common dictionary, or to any length of phrase in any given vocabulary. He was the first, who made this application of figures to telegraphic communication.

Much has been urged by various modern claimants for the honor of the invention of the telegraph. In England, the claims of Dr. Hooke, and of the Marquis of Worcester, to the original idea, are incontestable. But the invention long lay dormant, till wakened into active service by the French. Long before the French telegraph appeared, my father had tried his first telegraphic experiments. As he mentions in his own narrative*,

* Vol. i., page 145.

he tried the use of windmill sails in 1767 in Berkshire; and also a nocturnal telegraph with lamps and illuminated letters, between London and Hampstead. He refers for the confirmation of the facts to a letter of Mr. Perrot's, a Berkshire gentleman, who was with him at the time. The original of this letter is now in my possession. It was shewn in 1795 to the president of the Royal Irish Academy. The following is a copy of it:

"DEAR SIR,

"I PERFECTLY recollect having several conversations with you in 1767 on the subject of a speedy and secret conveyance of intelligence. I recollect your going up the hills to see how far, and how distinctly, the arms (and the position of them) of Nettlebed windmill sails were to be discovered with ease.

"As to the experiments from Highgate to London, by means of lamps, I was not present at the time, but I remember your mentioning the circumstance to me in the same year. All these particulars were brought very strongly to my memory when the French, some years ago, conveyed intelligence by signals; and I then thought, and declared, that the merit of the invention undoubtedly belonged to you. I am very glad that I have it in my power to send you this confirmation, because I imagine there is no other person now living, who can bear witness to your observations in Berkshire.

"I remain, dear Sir,

"Your affectionate friend,

"*Bath, Dec.* 9, 1795. "JAMES L. PERROT."

Claims to priority of invention are always listened to with doubt, or, at best, with impatience. To those who bring the invention to perfection, who actually adapt it to use, mankind are justly most grateful; and to these, rather than to the original inventors, grant the honors of a triumph. Sensible of this, the matter is urged no farther, but left to the justice of posterity.

I am happy to state, however, one plain fact, which stands independent of all controversy, that my father's was the *first*, and, I believe, the only telegraph, which ever spoke across the channel from Ireland to Scotland. "He was," as he says in his essay on this subject, "ambitious of being the first person, who should connect the islands more closely by facilitating their mutual intercourse;" and on the 24th of August, 1794, my brothers had the satisfaction of sending, by my father's telegraph, four messages across the channel, and of receiving immediate answers, before a vast concourse of spectators. The following lines were written on the occasion. To the numerical figures which *spoke* the first line, the pointers of his telegraph remain for ever fixed in their engraved representation in the Memoirs of the Royal Irish Academy.

"Hark! from basaltic rocks and giant walls,
To Britain's shores the glad Hibernia calls:

Her voice no longer waits retarding tides,
The meeting coasts no more the sea divides.
—Quick, at the voice of fortune or of fame,
Kindles from shore to shore the patriot flame.
Hovering in air, each kindred genius smiles,
And binds in closer bands the sister isles."

Some human partiality might be permitted its master in favor of a telegraph, which had performed at least what no other has hitherto accomplished. Yet far from querulously insisting upon the adoption of his own, my father concludes all he says upon the subject with declaring, that he does not pretend to assert, that the means of telegraphic communication which he has devised are the best.—" Imitations without end," says he, " may be attempted; pointers of various shapes and materials may be employed; real improvements will also probably be made, and perhaps new principles may be adopted. The varieties of art are infinite, and none but persons of narrow understanding, who feel a want of resources in their own invention, are jealous of competition, or are disposed to monopolize discoveries. The thing itself must sooner or later prevail; for utility convinces and governs mankind; and however inattention or timidity may for a time impede its progress, I will venture to predict, that it will at some future period be generally practised, not only in these islands, but that it will, in time, become a means of communication between the most distant parts of the world, wherever arts and sciences have civilized mankind."

If the really private letters of individuals, written at the moment when the persons were warmly engaged in any public affair, could be seen, how often would they be in shameful contradiction to the ostensible motives and character of the writers. I am proud to show, that my father spoke the same sentiments in public and private. The following note was written to his wife, in the first moment of disappointment.

TO MRS. EDGEWORTH.

" *Dublin, Tuesday night,*

" MY DEAR WIFE, *Nov.* 14, 1796.

" AFTER waiting long, as is usual, and indeed, from the multiplicity of suitors, as is necessary, I saw the secretary this afternoon.

" The Admiralty in England do not chuse to extend their telegraphic fame; and the government here does not chuse to extend mine. The secretary is to write to me, but I know it is to convey a negative.

" My mind keeps its own place, and my happiness depends on you and yours, and not on them and theirs.

" I hope, if I do not escape from Dublin so as to reach

you before this letter can, to see you and my dear home on Saturday.

" Yours, I think I may say *ever*,

" RICHARD LOVELL EDGEWORTH."

The next letters to his intimate friend will shew with what ease and good humour he turned his attention immediately to fresh objects.

TO DR. DARWIN.

" *Edgeworth-Town,*
Dec. 18, 1796.

" MY DEAR DOCTOR,

" I TOOK up a sheet of paper above a month ago, upon which I wrote half a page to you, intending to fill it with a pompous account of my having established the most complete system of intelligence all through the kingdom, that ever was attempted by ingenuity, and accomplished by industry. But before the moon had completed her lunatic gyration, I found that I had been counting upon the faith of Lord Lieutenants, and their satellites. After having had a great deal of trouble and expense, and some entertainment and improvement, I was turned off to graze. Whether my independence, or disinterestedness, or some other crime, was the cause, I cannot tell; but I have dared to publish the whole transaction *pro bono publico*, believing, that if men would abstain from abuse or calumny, and publish facts under their own names, it would be of use to those who govern, and to those who are to be governed. I have besides, in my own appeal, touched upon some general subjects of Irish politics. Mr. Johnson will send you a copy as soon as it is printed, but I do not insist upon your reading it.

" In one of your letters some time ago, you advised us to read Dugald Stewart, and to write upon education. Stewart we have read with great profit and pleasure, and we are writing upon education. Maria recurs frequently to your authority in a chapter on " attention," and has, I think—pardon my paternal partiality, managed your gigantic weapons with as much adroitness, as could be expected from a dwarf. Your new terms in Zoonomia require to be mouthed frequently to make them familiar; and in conversation we sometimes forget our grammar. She would write to ask you some questions, if she dared.

" You compose in your chaise, and I on horseback, which—there are certainly no other causes—is the reason why your lines roll so smoothly, and mine partake so much of tolutation. The verses at the end of Zoonomia I requested my son Sneyd (ten years old) to translate; but before I returned from my morning's ride, I had *done them* into English. Here they are.

The Triumphal Chariot of Medicine.

" The victor comes! I see his arms from far,
And hear the thunder of Hygeia's car.
Phœbus with laurel binds his brow, and Fame
Sounds from her silver car his deathless name.
Hurrying behind rides Age, with feeble cry,
Eager to tell the sage, that he must die.

" As I have presumed to hint an idea of little Sneyd's translating your verses, it is necessary to shew you, that he is not quite unfit for such a task. I must premise, that he had never seen Dryden's translation of Ovid, when he thus translated

The House of Sleep.

" Far in a vale there lies a cave forlorn,
Which Phœbus never visits, eve nor morn.
The misty clouds inhale the pitchy ground,
And twilight lingers all the vale around.
No watchful cock Aurora's beams invite;
No dogs, nor geese, the guardians of the night:
No flocks nor herds disturb the silent plains;
Within the sacred walls mute quiet reigns,

And murmuring Lethé soothing sleep invites.
——— In dreams again the flying past delights,
From milky flow'rs, that near the cavern grow,
Night scatters the collected sleep below."

" I find my paper growing *scant*, and I had a great many things to talk about.—I am to thank you for saving the eyes of a valuable horse. He was young; I had bred him; and till I read Zoonomia, I could not account for the frequent inflammation of his eyes. He and two other horses were going blind. From the construction of the stable, and the trimming of the inside of their ears, I suspected that what you had said was applicable to them *. Since I put caps on the horses' ears, they have got well.

" In the 'Sea-sick Minstrel,' page 50, there is a curious physico-medical anecdote. 'A young man, under the gripe of the tiger, appeared to move his lips, but uttered no sound. When he got free, he reproached the bystanders with not having answered to his screams.'

" See 'Journal Etranger' for 1756, page 55, for an anecdote about physicians.

" Adieu, my dear Doctor,

" R. L. E."

FROM MR. EDGEWORTH TO DR. DARWIN.

" *Edgeworth-Town,* 1796.

" VARIOUS events have disturbed the noiseless tenor of my way since I last wrote to you, my dear Doctor. Did I tell you, that my memorial to the Lord Lieutenant was rejected?

" He refused to listen to my proposals for establishing a communication by telegraphs across the water from London to Dublin in a few minutes; and oh! grievous mortification, I see that a telegraph is now at work from London to Deal and Dover.

" I had also proposed myself a candidate for the county,

* See Zoonomia, vol. ii, p. 237.

and for one whole month had no opponent, and have, notwithstanding, been obliged to relinquish.

"I believe you can guess, that I applied with activity to these pursuits, whilst I was engaged in them; but you will not so readily believe, that I bear these disappointments, not like an Epicurean, as some people think me, but as if I had been bred in the Stoic school of Mr. Day and Dr. Small. Whilst I was in the midst of my canvas, I wrote verses: not very good ones, but easy enough to shew that my mind was at ease. Perhaps it will not add to your ease to read them. I will spare you half—

"Four coursers bear me through the ranks of war,
Eight steeds successive whirl my rapid car;
From town to town, from house to house I fly,
Yet where's our candidate? the voters cry,——
So, from each corner of some festive hall,
At merry Christmas, eager children call;
Still in the middle stands the fool confest,
By all invited, and of all the jest.—
Vain all my toils! Lord **-***-**** appears,
Loud shakes his purse in every voter's ears;
And agents bribe, with promises and lies
Point to the Church, the Army, the Excise.—
Can poverty from gold withdraw his hand?
A gauger's rod, what vote can withstand?
Retire, presumptuous man, in time retire!
Say, if thou canst, to what thou wouldst aspire?
With friendship, love, and philosophic ease,
Form'd to be pleas'd, and wishing still to please—
Say, couldst thou add one real pleasure more,
To all the blessings you enjoy'd before?—
Couldst thou retard by all that man could say
Thy country's ruin, for one single day?—
Retire, presumptuous man, in time retire,
Leave knaves to plunder, and let fools admire!

" You will approve, my dear Doctor, the philosophy of these lines, though you may condemn the poetry.

" Your protegé, Lovell, is at Edinburgh, his own master, —he bears the northern blasts stoutly,

" Breathes the keen air, and carols as he goes."

" I intend to visit him at Edinburgh in the course of the summer. I shall know more of his inclination and future destination there, where he has made acquaintance of his own, than I could do at home. I wish him to mix with the world as much as may be, to acquire experience of others, and confidence in himself. Hitherto I have reason to be pleased. I intend to put him into *entire* possession of an estate of some hundreds a year, with a pretty house on it; as he disposes of this, I shall be able to judge how he will dispose of more.

" We are now preparing a didactic work on education. I seriously think, that more good may be done by improving education, than by any other means; and to that we apply our feeble powers with strong good will. I have lately endeavoured to excite a clergyman in this neighbourhood, who writes excellent prose and tolerable verse, to employ his pen, which has never yet decreased his poverty, in cutting away all that is useless, indecent, and absurd, from Herodotus, Plutarch, &c. I mean from some of the translations of them for young people.

" A pantheon taken from ancient gems has long been a favourite object of my wishes, in which Venus shall appear, not exactly like a street-walker; and Hermes, something better than George Barrington, &c. But alas! there is but one pencil in the world, that could draw any thing tolerably fit for the temple which I want to open to my pupils; and he has given to the vegetable world what was claimed for the mansions of the gods. But to return to my poor clergyman: I have advised him to select from Dryden, Pope, &c.; to fine-draw the pieces together with his best skill, to illustrate his Pantheon.—I have offered money to help him forward.

" What say you to my project?

"Perhaps the wheel of Fortune may project me from its rim into your neighbourhood this summer. If it does, nothing shall prevent me from paying you a visit.

"Adieu, yours most sincerely,

"R. L. Edgeworth."

Though he was unsuccessful in his canvas of the county of Longford at this time, my father saw, that he had so strong an interest, and was so much desired as a representative by the independent electors, that he had full encouragement on a future occasion to press his claims.

In the succeeding year, all his active pursuits were suspended. Mrs. Edgeworth's health, which had long been precarious, rapidly declined: consumption, which had been dreaded for her from early youth, had, by unremitting care and medical skill, been averted longer than those, who best knew the delicacy of her constitution, had thought possible. All of my father's letters to Dr. Darwin, that relate merely to her health, are of course suppressed, but I will not omit the following.

MR. EDGEWORTH TO DR. DARWIN.

"My dear Doctor, I wish you could infuse a fresh portion of health into the constitution of my poor wife.—But I fear, alas! that is impossible * * * * * She declines rapidly. But her mind suffers as little as possible.—I am convinced from all that I have seen, that *good sense* diminishes all the evils of life, and alleviates even the inevitable pain of declining health. By good sense, I mean that habit of the understanding, which employs itself in forming just estimates of every object that lies before it, and in regulating the temper and conduct. Mrs. Edgeworth, ever since I knew her, has carefully improved and cultivated this faculty; and I do not think I ever saw any person extract more good, and suffer less evil, than she has, from the events of life * * * * * * . She has not, and she never has had, that restless desire of changing place, which is so common to her situation. On the contrary, she is fearful of change, or of leaving home. You saw the patience of her sister.—Hers is not in the least inferior. We must all die,—we all know it. But to see those, whom we love, suffer and fade from day to day, is a cruel situation.—I intended to have written on other subjects, but this prevents me, * * * * * * "

Mrs. Edgeworth died in the Autumn of the year 1797.

I have heard my father say, that during the seventeen years of his marriage with this lady, he never once saw her out of temper, and never received from her an unkind word, or an angry look. Her solicitude and attention in the education of a large family of children were unremitting, greater than her health could bear, and such as even maternal affection would have found difficult, perhaps impossible to sustain, unless they had been supported by attachment to a husband of superior mind.

CHAPTER VIII.

[1797.

DURING the fifteen preceding years of which I have given an account, the variety of my father's employments never prevented him from attending to his great object—the education of his children. On the contrary, the variety of his occupations assisted in affording him daily and hourly opportunities for giving instruction after his manner, without formal lectures or lessons. For instance, at the time when he was building or carrying on experiments, or work of any sort, he constantly explained to his children whatever was doing or to be done; and by questions adapted to their several ages and capacities, exercised their powers of observation, reasoning, and invention.

It often happened, that trivial circumstances, by which the curiosity of the children had been excited, or experiments obvious to the senses, by which they had been interested, led afterwards to deeper reflection o to philosophical enquiries, suited to others in the family of more advanced age and knowledge. The animation spread through the house by connecting children with all that is going on, and allowing them to join in thought or conversation with the grown-up people of the family, was highly useful, and thus both sympathy and emulation excited mental exertion in the most agreeable manner.

In trying experiments he always shewed, that he was intent upon learning the truth, not upon supporting his opinion. By the examples he thus set us of fairness, candor, and patience, he trained the understanding to follow the best rules of philosophising; and, what is of more consequence for the happiness of the individual, he taught his pupils to apply philosophy to the government of the temper.

He explained, and described clearly,

> " With words succinct, yet full without a fault,
> He said no more than just the thing he ought."

This is as good a description of a judicious preceptor, as of a great orator.—He knew so exactly the habits, powers, and knowledge of his pupils, that he seldom failed in estimating what each could comprehend or accomplish. He saw at once where their difficulty lay, and knew how far to assist, how far to urge the mind, and where to leave it entirely to its own exertions. His patience in teaching was peculiarly meritorious, I may say surprising, in a man of his vivacity. He would sit quietly while a child was thinking of the answer to a question, without interrupting, or suffering it to be interrupted, and would let the pupil touch and quit the point repeatedly; and without a leading observation or exclamation, he would wait till the steps of reasoning and invention were gone through, and were converted into certainties. This was sometimes trying to the patience of the by-standers, who often decided, that the question was too difficult; when just at the moment that the silence and suspense could be no longer endured, his judgment has been justified, and his forbearance rewarded, by the child's giving a perfectly satisfactory answer.

The tranquillizing effect of this patience was of great advantage. The pupil's mind became secure, not only of the point in question, but steady in the confidence of its future powers. It was his principle to excite the attention fully and strongly for a short time, and *never to go to the point of fatigue.* But I do not assert, that he always adhered to this principle: he was so anxiously interested for each step in the progress of his pupils, and they were so eager to obtain his approbation, that they often, without his or their perceiving it, persevered too long. No wonder! for his praise was so delightful!—it was so warmly, so fondly given.

The cool by-standers might have thought, that this would have inspired vanity; but against this danger there was a preservative, even in the manner in which that praise was given; there was mixed with it so much affectionate sympathy, so much parental triumph, in his children's success, that affection for him, more than vanity for themselves, was excited by his applause, and from experience they insensibly drew the conclusion, that affection is better worth having than admiration.

In the education of the heart, his warmth of approbation, and strength of indignation, had powerful and salutary influence in touching and developing the affections. The scorn in his countenance, when he heard of any base conduct; the pleasure that lighted up his eyes, when he heard of any generous action; the eloquence of his language, and vehemence of his emphasis, commanded the sympathy of all who could see, hear, feel, or understand. Added to the power of every moral or religious motive, sympathy with the virtuous enthusiasm of those we love and reverence produces a great and salutary effect.

It often happens, that a preceptor appears to have great influence for a time, and that this power suddenly dissolves. This is, and must be the case, wherever any sort of deception has been used. My father never used any artifice of any kind, and consequently he always possessed that confidence, which is the reward of plain dealing; a confidence which increases in the pupil's mind with age, knowledge, and experience. I dwell on this reflection, certainly, with pride and pleasure, as far as it concerns my father and my beloved preceptor; but independently of private feelings, I trust, that my strong assertion of this fact may be useful to the public. It may tend to convince parents, that permanent influence over their children, that that influence which arises from grateful esteem, that which alone can endure from youth to age, may with certainty be obtained by PLAIN TRUTH.

The account of the manner, in which my father educated his family, naturally leads me to what he wrote on education. He never recommended any thing to the public, till he had put it to the test of experience. Long before he ever thought of writing or publishing, he had kept a register of observations and facts, relative to his children. This he began in the year 1798. He and Mrs. Honora Edgeworth kept notes of every circumstance which occurred worth recording, Afterwards Mrs. Elizabeth Edgeworth and he continued the same practice, and in consequence of his earnest exhortations, I began in 1791 or 1792, to note down anecdotes of the children, whom he was then educating. Besides these, I often wrote for my own amusement and instruction some of his conversation-lessons, as I may call them, with his questions and explanations, and the answers of the children. At the time these were taken, the idea of publishing any of them had never occurred.

To produce before the public, in a serious work, a number of apparently trivial anecdotes of children, even from infancy upwards, was an undertaking so obvious to commonplace mockery, and to the imputation of parental and family egotism, that it required for its execution his decision and strength of mind. To all who ever reflected upon education, it must have occurred, that facts and experiments were wanting in this department of knowledge, while assertions and theories abounded*. I claim for my father the merit

* This has been pointed out by a foreign critic of great

of having been the first to recommend, both by example and precept, what Bacon would call the experimental method in education.

---

ability. In speaking of Practical Education, he says, " this is the first time, that the idea of making education an experimental science has ever been developed. This is not a mere theory; the author traces the progress and trial of his experiments upon his pupils. All the merit, which depends * * * * * on long practice, may be ascribed to this work. The author has opened a new path, which may be pursued by men of ingenuity and observation, to the great advantage of mankind.

" Observation and experience, which have so much advanced our knowledge in physics, may, perhaps, with equal success, be applied to the science of education. By observing each step in the progress of the human mind, we may hope to succeed in laying the foundation of wise theory on accurate knowledge. Parents, directed by the ideas of this author, may endeavor to collect a succession of facts upon the physical and moral progress of their children, upon the influence of climate and of regimen on their health, the effect of external circumstances, and of all the casual events, which may come within the power of observation. We may, perhaps, by a comparison of a number of similar facts, obtain means of discovering general laws, which have not yet been determined, for the developement of the human mind. Rules, by which probabilities may be calculated with a precision hitherto unknown.

" If this be but a vain hope—if the length of the periods of observation—if the variety, uncertainty, and infinite complication of facts, render their results vague and imperfect, still some truths will probably be discovered useful to parental prudence and solicitude. This good will result from the awakened spirit of observation; that parents will be more attentive to their children; that the interest of curiosity will be added to

If I were obliged to rest on any single point, my father's credit as a lover of truth, and his utility as a philanthropist and a philosophical writer, it should be on his having made this first record of experiments in education. The example, which he has set, has been followed in some families, and will be followed in others, by parents who are really anxious to know and to improve the dispositions of their children, and to instruct themselves in the methods of cultivating their understandings and hearts. I must here reiterate a caution, which my father has given elsewhere.

In noting anecdotes of children, the greatest care must be taken, that the pupils should not know, that any such register is kept. Want of care in this particular would totally defeat the object in view, and would lead to many and irremediable bad consequences; would make the children affected and false; or would create a degree of embarrassment and constraint, which must prevent the natural action of the understanding or the feelings. Nothing could be devised more pernicious, both to parents and children, than

---

the natural tenderness of their feelings; and that ties, which, for the happiness of mankind, cannot be drawn too close, will be rendered at once more strong and more engaging."

opening a repository for infant wit, or introducing a spirit of ostentation and competition among pupils and their preceptors. Those who do not feel, that they are above such motives, and that they are capable of the degree of forbearance and discretion necessary to the purpose, are here implored, never to attempt such a plan, either for their own private use, or for the public. It would worse than fail in their hands; their children would be spoiled, and the public would be deluged with nonsense.

On the other hand, when these precautions are taken, it is to be hoped, that parents and preceptors will not suffer themselves to be deterred by the cowardly fear of ridicule, from noting down as many *facts* relative to early education as can be obtained. In the registry of such observations, considered as contributing to a history of the human mind, nothing should be rejected as trivial. The circumstances, which may seem most trifling to vulgar observers, may be most valuable to the philosopher; they may throw light, for example, on the manner in which ideas and language are formed and generalised. From a collection of facts honestly stated, future metaphysicians and writers on the philosophy of the mind might select new data for their reasonings, and new means of awakening the attention of their readers.

Parents of different habits and views, and in various classes of society, have mentioned to us the anecdotes of children in Practical Education as having given them a peculiar interest and confidence in that book. These gave internal evidence, that it could not have been composed by a person who was unacquainted with children, or who was uninterested about them, and intent only on stringing fine sentences together. It was obvious, that the plan had at least been tried for a number of years. The father might be mistaken in many things; but still it was evident, that he was a real father, eagerly in earnest in the education of his own children. All through Practical Education, as it was written by two people, the words *we* and *our* are frequently used.

"*We* think so and so."—"*Our* pupils do so and so."

Without circumlocution fatiguing to the reader, and awkward to the writer, this could not have been avoided; but I felt and expressed repugnance to using the words *our* pupils, because I knew it would lead to the false conclusion, that I educated the children of whom we spoke, and the fact was not so. Only one child (a beloved brother, since dead) was, during the early years of his life, entrusted to my care. My Charlotte was under the excellent and successful care of my sister Emmeline (now Mrs. King). Another sister, Honora, named after her whom we lost, was, from the time she was six years old, under the care of Mrs. Charlotte Sneyd, her aunt:—all these children were of course under the superintending guidance of my father and of Mrs. Edgeworth, the lady mentioned in the preface to Practical Education.

It was my father's delight to say, that, in literature, his thoughts and mine were in common; he never would permit me to attribute to him even what was peculiarly his own. In the work, of which I am now speaking, the principles of education were peculiarly his, such as I felt he had applied in the cultivation of my own mind, and such as I saw in the daily instruction of my younger brothers and sisters during a period of nearly seventeen years; all the general ideas originated with him, the illustrating and manufacturing them, if I may use the expression, was mine. This first partnership work (Practical Education) was published in 1798.

So commenced that literary partnership, which, for so many years, was the pride and joy of my life.

CHAPTER IX.

My father was past fifty, when he was left a third time a widower, with a numerous family, by different wives: four sons and five daughters living with him; some of them grown up, others very young—the youngest but three years old—two of the daughters fourteen and sixteen, just at the age when a mother's care is of most importance. Besides his children, two sisters of the late Mrs. Edgeworth had resided with us for several years. Though they had friends and near connexions in England, for whom they felt high esteem, they had remained in Ireland with us, and they formed part of this large family, attached to them by ties of kindred, and by feelings of gratitude and esteem. Those who knew him intimately, and all indeed who had seen how much the felicity of his former life had depended upon conjugal affection, were aware, that he would not be happy unless he married again.

The life of every man, as it has been well observed, "is a continual chain of incidents, each link of which hangs upon the former." Even where no striking incidents appear, the chain of causation may often be traced to some habit, passion, or taste, of the individual; it is curious to observe how differently the fates of men are decided, some by the occurrence of accidental circumstances, others by internal causes, constantly recurring in their mind, and belonging to their characters: those governed by the will of others, these by their own decision.

After the first years of youth, my father's destiny in life was never decided by what is called accident; nor was he drawn against his better judgment by the persuasion of others. His own principles, his sense of duty, and the conviction of his understanding, appear to have determined constantly his course of life; his discrimination of character, and his tastes for literature and science, uniformly directed him in the choice of his friends; these, and not accident, led to his forming those nearer and dearer connexions, on which his uncommon domestic happiness depended.

In his own narrative, he has shewn how much his destiny was influenced by his predominant taste for mechanics, which introduced him to Dr. Darwin, led him to Lichfield, and to his acquaintance with Miss Seward, and with Honora Sneyd. The same tastes, still prevailing, prepared and influenced in a similar manner his choice in another connexion, and decided his fate at a much later period of life.

Many years previous to this time, in the summer of 1774, when he was just married to Miss Honora Sneyd, in the bridal visit which they paid at Black Castle to my father's favorite sister, Mrs. Ruxton, they met at her house Dr. Beaufort, his wife, and daughter. Dr. Beaufort's name is well known to the British public as the author of our best map of Ireland, and most valuable Memoir on the Topography, and Civil and Ecclesiastical State of this country. He is still better known in Ireland as an excellent clergyman, of a liberal spirit and conciliating manners, and as a man of taste and literature. My father was much pleased with him, and desired to cultivate his acquaintance. The daughter, who was with Dr. Beaufort on this visit, was a little child of six years old, in a white frock and pink sash: her image was fixed in my father's recollection by a question that occurred, whether her mother did or did not spoil her? He could little foresee how much influence this child was to have, years afterwards, on his happiness.

Dr. Beaufort and his family went to reside in England, and my father never saw him for many years. In 1785 their acquaintance was again renewed by meeting at the Royal Irish Academy, and at the house of Dr. Ussher (Professor of Astronomy to the University of Dublin). Dr. Beaufort had received from Edmund Burke a commission to procure the skeleton of a moose deer; the bones of these deer were to be found in a lake near Edgeworth-Town, and my father undertook the commission for the Doctor. This led to a slight correspondence, advancing their intimacy a degree or two. A year afterward Dr. Beaufort, in a progress through Ireland, to collect materials for his civil and ecclesiastical map, paid us a visit of a few days at Edgeworth-Town; but for subsequent years, living at a distance, we saw nothing of him. Meantime, Dr. Beaufor 's little daughter grew up. My father's sister, Mrs. Ruxton, whose intimacy with the Beaufort family continued, was particularly fond of Miss Beaufort, and often spoke of her to us as a friend and companion peculiarly suited to her taste; still we never met. The acquaintance was renewed, when my father went to see his friend Mr. Foster, at Collon, of which place Dr. Beaufort was vicar.

It happened that Miss Beaufort, who possessed uncommon talents for drawing, had, at the request of my aunt Ruxton, sketched some designs for "The Parent's Assistant," which were shewn to my father, and which he criticised as freely, as though they had not been the work of a lady, and designed for his daughter. He was charmed with the temper and good sense, with which these criticisms were received. But this impression, however favourable, was but slight. They were again separated, and it was not till some time afterwards, when in a visit, which she, with her family, paid at Edgeworth-Town in 1798, my father had an opportunity of discerning, that she possessed exactly the temper, abilities, and disposition, which would ensure the happiness of his family as well as his own, if he could hope to win her affections.

The marked approbation of a man of distinguished abilities could not but be gratifying to the object of his attention. She had conversed with him with perfect freedom, admiring the various knowledge and genius, which appeared in his conversation; attentive to all that could improve her own understanding, or assist her taste for literature; listening to his counsels as to those of a friend, but obviously without a thought of him as a lover. The difference of their age prevented her suspecting his attachment, till a few hours before it was declared. The thought of his large family, more than any disparity of age, was the first and great objection. My father's estate, though considerable, was not sufficient to afford any settlement, that could have been a temptation in the Smithfield-bargain way of estimating things. If the lady's parents had not been far above such views, they could scarcely have been justified in advising the connexion. They neither pressed nor dissuaded, but left their daughter entirely to her own judgment and inclinations.

When I first knew of this attachment, and before I was well acquainted with Miss Beaufort, I own that I did not wish for the marriage. I had not my father's quick penetration into character; I did not at first see the superior abilities, or qualities, which he discovered; nor did I anticipate any of the happy consequences from this union, which he foresaw. All that I thought, I told him. With the most kind patience he bore with me, and instead of withdrawing his affection, honoured me the more with his confidence. He took me with him to Collon, threw open his whole mind to me—let me see all the changings and workings of his heart. I remember his once saying to me, "I believe, that no human creature ever saw the heart of another more completely without disguise, than you have seen mine." I can never, without the strongest emotions of affection and gratitude, recollect the infinite kindness he shewed me at this time, the solicitude he felt for my happiness at the moment when all his own was at stake, and while all his feelings were in the agony of suspense; the consequence was, that no daughter ever felt more sympathy with a father, than I felt for him; and assuredly the pains he took to make me fully acquainted with the character of the woman he loved, and to make mine known to her, were not thrown away. Both her inclination and judgment decided in his favor. His eloquent affection conquering her timidity, and inspiring her with the necessary and just confidence in her own abilities, she consented to undertake the great responsibility of becoming the mistress of that large family, of whose happiness she was now to take charge.

The manner in which he announces his intended marriage to Dr. Darwin at the end of the following letter, beginning with a variety of indifferent subjects, is characteristic.

"MY DEAR DOCTOR, 1798.

"I have received from England a very fine head of you:—when I say a *fine* head, I mean only as to the execution; for it does not give an idea of any faculty of your mind, except ingenuity:—there is a cloud over your brow, and a compression of the lips, that hide your benevolence and good humour And great author as you are, my dear Doctor, I think you excel the generality of mankind as much in generosity, as in abilities. I observe with pleasure, mixed with pride, the rapid growth of your fame. We wait with impatience for your new book. We have wished to collect some facts for you; but before I finish my letter you will see, that *I* had other game in view. I must give you my notes in disorder.

" The Upas, for which you have been so much abused, *may* not grow in Java:—but a traveller of undoubted veracity, *Stedman*, (Voyage to Surinam, vol. ii, p. 183,) mentions qualities of the marcoory tree almost as formidable, and not dissimilar.

" Frogs, and other small animals frozen, have their limbs as brittle as a tobacco-pipe. The broken frozen limbs heal, when gradually thawed, but if refrozen, they never coalesce. (Vid. Hearne's Hudson's Bay Discoveries, p. 397.) I suppose, that the second freezing breaks off the corresponding indentures of the fracture.

" Are you still bent upon agriculture? If you are, here is a note and query for you. In 1787 I rented a farm of thirty acres, at a rack rent of half a guinea per acre. I improved it, was never at any one time one hundred pounds out of pocket; and in five years had seventeen pounds clear profit to balance of accounts. I let it on a long lease, at one guinea and a half per acre, to a rich tenant.—Query: how was it improved? A most exact account was kept of profit and loss.

" I propose to heat hot-houses by pipes laid through dunghills, and communicating with the open air at one end, and opening into the house at the other. Such pipes, in common stable-dung, continue to give out air at a regular heat of 95° Fahrenheit for many days, as Lovell and I have tried in a pipe six inches diameter. I saw one of your old inventions claimed by somebody in the papers, the other day,—two balls chained together, and shot at once from separate cannon.

" The speaking machine, which is just announced from

France, does not say so many words as yours did many years ago. It prattles only papa and mamma. Yours spoke those words, and could also say GO. I placed one of your mouths in a room near some people in 1770, who actually thought I had a child with me, calling papa and mamma.

"And now for my piece of news, which I have kept for the last. I am going to be married to a young lady of small fortune and large accomplishments,—compared with my age, much youth (not quite 30), and more prudence—some beauty, more sense—uncommon talents, more uncommon temper,—liked by my family, loved by me. If I can say all this three years hence, shall not I have been a fortunate, not to say a wise man? "R. L. E."

While my father's domestic happiness seemed preparing so smoothly, public affairs in Ireland wore a stormy aspect. This was in the year 1798. A few days after the preceding letter was written, we heard, that a conspiracy had been discovered in Dublin, that the city was under arms, and its inhabitants in the greatest terror. Doctor Beaufort and his family were there. My father, who was at Edgeworth-Town, set out immediately to join them.

On his way he met an intimate friend of his; one stage they travelled together, and a singular conversation passed. This friend, who as yet knew nothing of my father's intentions, began to speak of the marriage of some other person, and to exclaim against the folly and imprudence of any man's marrying in such disturbed times, "no man of honor, sense, or feeling, would incumber himself with a wife at such a time!"—My father urged, that this was just the time, when a man of honor, sense, and feeling, would wish, if he loved a woman, to unite his fate with hers, and to acquire the right of being her protector.

The conversation dropped there. But presently they talked of public affairs—of the important measure expected to be proposed of a union between England and Ireland—of what would probably be said and done in the next session of Parliament; my father, foreseeing that this important national question would probably come on, had just obtained a seat in Parliament*. His friend, not knowing or recollecting this, began to speak of the imprudence of commencing a political career late in life.

"No man, you know," said he, "but a fool, would venture to make a first speech in Parliament, or to marry, after he was fifty."

My father laughed, and surrendering all title to wisdom, declared, that, though he was past fifty, he was actually going in a few days, as he hoped, to be married, and in a

---

* He was elected for the borough of St. John's Town, county of Longford.

few months would probably make his "first speech in Parliament."

His friend made as good a retreat as the case would admit, by remarking, that his maxim could not apply to one who was not going either to be married or to speak in public for the first time.

As fast as possible my father pursued his way to Dublin. He found the city as it had been described to him, under arms, in dreadful expectation. The timely apprehension of the heads of the conspiracy at this crisis prevented a revolution, and saved the capital. But the danger for the country seemed by no means over—insurrections, which were to have been general and simultaneous, broke out in different parts of the kingdom. The confessions of a conspirator, who had turned informer, and the papers seized and published, proved, that there existed in the country a deep and widely spread spirit of rebellion. Though disconcerted by the present vigilance or good fortune of government, it was hardly to be doubted, that fresh attempts, in concert with foreign enemies, would be made in future. From different reasons, all parties thought the country in a dangerous situation, and foresaw, that there must be further commotions.

Instead of delaying his marriage, which some would have advised, my father urged for an immediate day. On the 31st of May he was married to Miss Beaufort, by her brother, the Rev. William Beaufort, at St. Anne's church, in Dublin. They came down to Edgeworth-Town immediately, through a part of the country that was in actual insurrection. Late in the evening they arrived safe at home, and my father presented his bride to his expecting, anxious family.

Of her first entrance and appearance that evening I can recollect only the general impression, that it was quite natural, without effort or pretension. The chief thing remarkable was, that she, of whom we were all thinking so much, seemed to think so little of herself.

A more trying situation for a wife could hardly be imagined, than that in which she was now placed. She knew, that in the minds of all who surrounded her—sons, daughters, and sisters-in-law, old associations and present feelings, though not averse to her individually, must be painfully affected by the first introduction of a new wife and mother. She was aware, that points of comparison must continually recur with those, who had been much beloved or highly admired. Love and sorrow for their late mother were still fresh in the minds of her own children; while ever present to the memory of others of the family, and of traditional power over the imagination, was the character of one highly gifted and graced with every personal and mental endowment—the more than *celebrated,* the revered Honora! Knowing and feeling all this—and who could know or feel it more—my father seemed neither embarrassed nor anxious for his present wife; not imprudently impatient to have her admired or beloved by his family.

Soon after this marriage, things and persons found themselves in their proper places; and the fear of change, which had perplexed numbers, was gradually dispelled. The sisters of the late Mrs. Edgeworth, those excellent aunts (Mrs. Mary and Charlotte Sneyd), instead of returning to their English friends and relations, remained at Edgeworth-Town. This was an auspicious omen to the common people in our neighbourhood, by whom they were universally beloved—it spoke well, they said, for the *new* lady. In his own family, the union and happiness she would secure was soon felt, but her superior qualities, her accurate knowledge, judgment, and abilities, in decision and in action, appeared only as occasions arose and called for them. She was found always equal to the occasion, and superior to the expectation. The power and measure of her efficient kindness could never be calculated, and was never fully known to each individual of her family, till by that individual it was most wanted.

This lady, thank God, is still living!—and, thank God, still living with us. No one can disdain flattery more than she does, or than I do. It is unworthy of her and of myself. She will see this before it is printed, and I am aware, that, though she will be certain that I think and feel what I say, she will at first wish, that this page should be suppressed; but I claim from her affection to my father the right to state opinions and facts necessary to do justice to his judgment and his character—essential to prove, that he did not late in life marry merely to please his own fancy, but that he chose a companion suited to himself, and a mother fit for his family. This, of all the blessings we owe to him, has proved the greatest.

CHAPTER X.

The summer of 1798 passed without any interruption of our domestic tranquillity. Though disturbances in different parts of Ireland had broken out, yet now, as in former trials, the County of Longford remained quiet; free at least from open insurrection, and, as far as appeared, the people well disposed. They complained, however, very frequently to my father of the harassing of certain new-made justices of the peace, and yeomen military, or, as the people called them, *scourers of the country*, who, galloping about night and day, would let no poor man sleep in peace. Our magistracy had at that time fallen below its proper level; many of the great proprietors of this county were absentees; and for want of resident gentlemen, magistrates were made of men without education, experience, or hereditary respectability. During the war, and in consequence of what were called the *war-prices*, graziers, land-jobbers, and middle-men had risen into comparative wealth; and instead of turning in due season, according to the natural order of things, into Buckeens and Squireens, they had been metamorphosed into justices of the peace and committee men, or into yeomen lieutenants and captains. In these their new characters, they bustled and bravadoed; and sometimes from mere ignorance, and sometimes in the certainty of party support or public indemnity, they overleaped the bounds of law. Upon slight suspicion, or vague information, they took up and imprisoned many who were innocent; the relations of the injured appealed to him, who was known to be the friend of public justice. I will not say *the friend of the poor*, though this was the name by which I have often heard him called. But this has become a hackneyed expression, degraded from its real meaning, since it has been used for party purposes, or by those who aim only at vulgar popularity. In consequence of appeals to him, my father made inquiry at public sessions or assizes into various cases of persons, who had been imprisoned. Sometimes such examinations, warrants, and committals, were produced, or such explanatory letters were written to him by justices of the quorum, worded in such a blundering manner, so spelled, so scrawled, as to be almost illegible and quite incomprehensible. All this would have been ludicrous, had not the matter been too serious for ridicule, where the liberties and lives of human creatures were at stake. My father exerted himself upon all occasions to keep the law in its due course, representing, that, whether the accused were innocent or guilty, they were entitled to fair trial; that, till it was proved that they had forfeited the protection of our constitution, no persons should be treated as enemies or outlaws; that it was bad policy to make the people detest the authority, which they were bound to obey, and on their obedience to which the safety of all ranks depends.

Those who were conscious of making themselves objects of dislike to the lower class of people were naturally afraid, that, if any disturbance arose, they should be the first victims; and cowardice combined with party prejudice, to increase their violence. They disregarded my father's representations; as far as they dared, resented his interference; and were in unfeigned astonishment at his opposite course of conduct. They found, that, when he was to judge of any action, he never inquired whether it had been committed by a catholic, or a protestant; nor would he use, or permit to be used before him, either as a magistrate or as landlord, any of those party names, which designate and perpetuate party hatred. He would not even understand, that the term *an honest man*, pronounced with peculiar emphasis, meant, with one party, exclusively a *protestant.* The principle of "shew me the man, and I will shew you the law," he considered as the opprobrium of magistracy. No fears of the timid, or cries of the violent; no personal views, no danger for himself of misrepresentation or odium, could in any one instance prevail upon him, even for the plea of public safety, or the necessity of the times, to strain a single point of law or justice. He believed that justice, like honesty, of which in fact it is only a more enlarged description, is always the best policy. He adhered therefore to the straight rule of right, without entering into any of the obliquities of expediency. The maxim, that extraordinary times call for extraordinary measures, he considered to be a principle, dangerous as it is vague; because those, who fancy they obey a mysterious legislative *call,* are often hurried on merely by the suggestions of their own fears, or their own passions; no judgment remaining to decide upon what is ordinary, or extraordinary. He thought, that no partial commotion, nothing less than a general earthquake of the state, should be suffered to disturb the balance of even-handed justice: when this is overturned, nothing remains but the sword.

Towards the autumn of the year 1798, this country became in such a state, that the necessity for resorting to the sword seemed imminent. Even in the county of Longford, which had so long remained quiet, alarming symptoms appeared, not immediately in our neighbourhood, but within six or seven miles of us, near Granard. The people were leagued in secret rebellion, and waited only for the expected arrival of the French army, to break out. In the adjacent countries military law had been proclaimed, and our village was within a mile of the bounds of the disturbed county of Westmeath. Though his own tenantry, and all in whom he put trust, were as quiet, and, as far as he could judge, as well disposed as ever, yet my father was aware, from information of too good authority to be doubted, that there were disaffected persons in the vicinity.

Numbers held themselves in abeyance, not so much from disloyalty, as from fear, that they should be ultimately the conquered party. Those who were really and actually engaged, and in communication with the rebels, and with the foreign enemy, were so secret and cunning, that no proofs could be obtained against them; while the eagerness to gain information laid proprietors in the middle and higher classes open to treachery, and to double danger from informers, who often gave them false clews, to involve them still farther in darkness and error. The object of such persons being in general to gratify private malice, or to favor the escape of the guilty by turning suspicion upon the innocent*.

---

* How ingeniously cunning the lower Irish are in contriving concealments and modes of escape is well known in Ireland, to every one who has been out on any of these *rebel* or *defender* hunts. One instance, which may be new to the English reader, I found in an old letter of this time—1798.

"A Mr. Pallas, who lived at Growse-Hall, lately received information, that a certain defender was to be found in a lone house, which was described to him. He took a party of men with him in the night, and he got to the house very early in the morning. It was scarcely light. The soldiers searched, but no man was to be found. Mr. Pallas ordered them to search again, for that he was certain the man was in the house; they searched again, but in vain; they gave up the point, and were preparing to mount their horses, when one man, who had staid a little behind his companions, saw, or thought he saw, something move at the end of the garden behind the house. He looked

Previous to this time, the principal gentry in the county had raised corps of yeomanry; but my father had delayed doing so, because, as long as the civil authority had been sufficient, he was unwilling to resort to military interference, or to the ultimate law of force, of the abuse of which he had seen too many recent examples. However, it now became necessary, even for the sake of justice to his own tenantry, that they should be

---

and beheld a man's arm come out of the ground: he ran to the spot and called to his companions, but the arm had disappeared; they searched, but nothing was to be seen; and though the soldier still persisted in his story, he was not believed. 'Come,' cries one of the party, 'don't waste your time here looking for an apparition amongst these cabbage-stalks—go back once more to the house.' They went to the house, and lo! there stood the man they were in search of in the middle of the kitchen.

"Upon examination it was found, that from his garden to his house there had been practised a secret passage under ground: a large meal-chest in the kitchen had a false bottom, which lifted up and down at pleasure, to let him into his subterraneous dwelling.

"Whenever he expected the house to be searched, down he went; the moment the search was over, up he came; and had practised this with success, till he grew rash, and returned one moment too soon.

"I do not vouch for the truth of this story, I only tell it you for your entertainment."

put upon a footing with others, have equal security of protection, and an opportunity of evincing their loyal dispositions. He raised a corps of infantry, into which he admitted Catholics as well as Protestants. This was so unusual, and thought to be so hazardous a degree of liberality, that by some of an opposite party it was attributed to the worst motives. Many who wished him well came privately to let him know of the odium, to which he exposed himself—the timid hinted fears and suspicions, that he was going to put arms into the hands of men, who would desert or betray him in the hour of trial; who might find themselves easily absolved from holding any faith with a Protestant, and with one of a family, of whom the head, in former times, had been distinguished by the appellation of *Protestant Frank.* He thanked his secret advisers, but openly and steadily abided by his purpose. Suspicion he knew often produces the very evils it fears. Confidence he felt to be due to those, who had hitherto conducted themselves irreproachably, and who manifested at this moment, by every means in their power, strong attachment to him and to the government. These men had been exposed to some trial, and might be called upon to resist great temptations of passion and interest. On his own part, my father knew the risk he ran, but he braved it. Resident as he had long been in Ireland, his established character, his property, and his large family, afforded altogether so strong a pledge of his good intentions, that he considered himself as in a situation to set an example of that conduct, which he thought most advantageous for the country. The corps of Edgeworth-Town Infantry was raised, but the arms were, by some mistake of the ordnance-office, delayed. The anxiety for their arrival was extreme, for every day and every hour the French were expected to land.

The alarm was now so general, that many sent their families out of the country. My father was still in hopes, that we might safely remain. At the first appearance of disturbance in Ireland he had offered to carry his sisters-in-law, the Mrs. Sneyds, to their friends in England, but this offer they refused. Of the domestics, three men were English and Protestant, two Irish and Catholic; the women were all Irish and Catholic, excepting the housekeeper, an Englishwoman, who had lived with us many years. There were no dissensions or suspicions between the Catholics and Protestants in the family, and the English servants did not desire to quit us at this crisis.

At last came the dreaded news. The French, who landed at Killala, were, as we learned, on their march towards Longford. The touch of Ithuriel's spear could not have been more sudden or effectual, than the arrival of this intelligence, in shewing people in their real forms. In some faces joy struggled for a moment with feigned sorrow, and then, encouraged by sympathy, yielded to the *natural* expression. Still my father had no reason to distrust those, in whom he had placed confidence; his tenants were steady; he saw no change in any of the men of his corps, though they were in the most perilous situation, having rendered themselves obnoxious to the rebels and invaders, by becoming yeomen, and yet standing without means of resistance or defence, their arms not having arrived.

The evening of the day, when the news of the success and approach of the French came to Edgeworth-Town, all seemed quiet; but early the next morning, September 4th, a report reached us, that the rebels were *up* in arms within a mile of the village, pouring in from the county of Westmeath, hundreds strong. Such had been the tranquillity of the preceding night, that we could not at first believe their report.—An hour afterwards it was contradicted. An English servant, who was sent out to ascertain the truth, brought back word, that he had ridden three miles from the village on the road described, and that he had seen only twenty or thirty men with green boughs in their hats, and pikes in their hands, who said, "*that they were standing there to protect themselves against the Orange-men, of whom they were in dread, and who, as they heard, were coming down to cut them to pieces.*" This was all nonsense; but no better sense could be obtained. Report upon report, equally foolish, was heard, or at least uttered. But this much being certain, that men armed with pikes were assembled, my father sent off an express to the next garrison town (Longford), requesting the commanding officer to send him assistance for the defence of this place. He desired us to be prepared to set out at a moment's warning. We were under this uncertainty, when an escort with an ammunition cart passed through the village, on its way to Longford.—It contained several barrels of powder, intended to blow up the bridges, and to stop the progress of the enemy. One of the officers of the party rode up to our house, and offered to let us have the advantage of his escort. But, after a few minutes deliberation, this friendly proposal was declined. My father determined, that he would not stir till he knew whether he could have assistance; and as it did not appear as yet absolutely necessary that we should go, we staid—fortunately for us!

About a quarter of an hour after the officer and the escort had departed, we, who were all assembled in the portico of the house, heard a report like a loud clap of thunder. The doors and windows shook with some violent concussion; a few minutes afterwards, the officer galloped into the yard, and threw himself off his horse into my father's arms almost senseless. The ammunition cart had blown up, one of the officers had been severely wounded, and the horses and the man leading them killed; the wounded officer was at a farm-house on the Longford road, at about two miles distance. The fear of the rebels was now suspended, in concern for this accident. Mrs. Edgeworth went immediately to give her assistance; she left her carriage for the use of the wounded gentleman, and rode back. At the entrance of the village she was stopped by a gentleman in great terror, who, taking hold of the bridle of her horse, begged her not to attempt to go further, assuring her that the rebels were coming into the town. But she answered, that she must and would return to her family. She rode on, and found us waiting anxiously for her. No assistance could be afforded from Longford; the rebels were reassembling, and advancing towards the village; and there was no alternative, but to leave our home as fast as possible. One of our carriages having been left with the wounded officer, we had but one at this moment for our whole family, eleven in number. No mode of conveyance could be had for some of the female servants; our faithful English housekeeper offered to stay till the return of the carriage, which had been left with the officer; and as we could not carry her, we were obliged, most reluctantly, to leave her behind to follow, as we hoped, immediately. As we passed through the village, we heard nothing but the entreaties, lamentations, and objurgations of those, who could not procure the means of carrying off their goods or their families: most painful when we could give no assistance.

Next to the safety of his own family, my father's greatest anxiety was for his defenceless corps. No men could behave better than they did at this first moment of trial. Not one absented himself, though many, living at a distance, might, if they had been so inclined, have found plausible excuses for nonappearance. The bugle was not sounded to call them together, but they were in their ranks in the street the moment they had their captain's orders, declaring, that whatever he commanded they would do. He ordered them to march to Longford. The idea of going to Longford could not be agreeable to many of them, who were Catholics; because that town was full of those who called themselves—I would avoid using party names if I could, but I can no otherwise make the facts intelligible—who called themselves Orangemen, and who were not supposed to have favourable opinions of any of another religious persuasion. There was no reluctance shewn, however, by the Catholics of this corps to go among them. The moment the word *march* was uttered by their captain, they marched with alacrity.—One of my brothers, a youth of fifteen, was in their ranks; another, twelve years old, marched with them.

We expected every instant to hear the shout of the rebels entering Edgeworth-Town. When we had got about half a mile out of the village, my father suddenly recollected, that he had left on his table a paper, containing a list of his corps; and that, if this should come into the hands of the rebels, it might be of dangerous consequence to his men; it would serve to point out their houses for pillage, and their families for destruction. He turned his horse instantly, and galloped back for it. The time of his absence appeared immeasurably long, but he returned safely, after having destroyed the dangerous paper.

About two miles from the village was the spot, where the ammunition cart had been blown up; the dead horses, swollen to an unnatural bulk, were lying across the road. As we approached, we saw two men in an adjoining field looking at the remains of one of the soldiers, who had been literally blown to pieces. They ran towards us, and we feared, that they were rebels going to stop us. They jumped over the ditch, and seized our bridles; but with friendly intent. With no small difficulty they dragged us past the dead horses, saying, "God speed you! and make haste any way!" We were very ready to take their advice. After this, on the six long miles of the road from Edgeworth-Town to Longford, we did not meet a human being. It was all silent and desert, as if every creature had fled from the cabins by the road side.

Longford was crowded with yeomanry of various corps, and with the inhabitants of the neighbourhood, who had flocked thither for protection. With great difficulty the poor Edgeworth-Town infantry found lodgings. We were cordially received by the landlady of a good inn. Though her house was, as she said, fuller than it could hold, as she was an old friend of my father's, she did contrive to give us two rooms, in which we eleven were thankful to find ourselves. All our concern now was for those we had left behind. We heard nothing of our housekeeper all night, and were exceedingly alarmed: but early the next morning, to our great joy, she arrived. She told us, that after we had left her, she waited hour after hour for the carriage: she could hear nothing of it, as it had gone to Longford with the wounded officer. Towards evening, a large body of rebels entered the village.—She heard them at the gate, and expected that they would have broken in the next instant. But one, who seemed to be a leader, with a pike in his hand, set his back against the gate, and swore, that, " if he was to die for it the next minute, he would have the life of the first man, who should open that gate, or set enemy's foot within side of that place. He said the housekeeper, who was left in it, was a good gentlewoman, and had done him a service, though *she did not know him, nor he her.* He had never seen her face, but she had, the year before, lent his wife, when in distress, sixteen shillings, the rent of flax-ground, and he would stand her friend now.

He kept back the mob; they agreed to send him to the house with a deputation of six, *to know the truth,* and to ask for arms. The six men went to the back-door, and summoned the housekeeper; one of them pointed his blunderbuss at her, and told her, that she must fetch all the arms in the house; she said she had none. Her champion asked her to say if she remembered him—" No; to her knowledge she had never seen his face." He asked if she remembered having lent a woman money to pay her rent of flax-ground the year before? " Yes," she remembered that, and named the woman, the time, and the sum. His companions were thus satisfied of the truth of what he had asserted. He bid her not to be *frighted,* " for that no harm should happen to her, nor any belonging to her; not a soul should get leave to go into her master's house; not a twig should be touched, nor a leaf harmed." His companions huzzaed and went off. Afterwards, as she was told, he mounted guard at the gate during the whole time the rebels were in the town.

When the carriage at last returned, it was stopped by the rebels, who filled the street; they held their pikes to the horses and to the coachman's breast, accusing him of being an Orange-man, because, as they said, he wore the orange colours (our livery being yellow and brown). A painter, a friend of ours, who had been that day at our house, copying some old family portraits, happened to be in the street at that instant, and called out to the mob, "*Gentlemen, it is yellow!—gentlemen, it is not orange.*" In consequence of this happy distinction they let go the coachman; and the same man, who had mounted guard at the gate, came up with his friends, rescued the carriage, and surrounding the coachman with their pikes brought him safely into the yard. The pole of the carriage having been broken in the first onset, the housekeeper could not leave Edgeworth-Town till morning. She passed the night in walking up and down, listening and watching, but the rebels returned no more, and thus our house was saved by the gratitude of a single individual.

We had scarcely time to rejoice in the escape of our housekeeper, and safety of our house, when we found, that new dangers arose even from this escape. The house being saved created jealousy and suspicion in the minds of many, who at this time saw every thing through the mist of party prejudice. The dislike to my father's corps appeared every hour more strong. He saw the consequences, that might arise from the slightest breaking out of quarrel. It was not possible for him to send his men, unarmed as they still were, to their homes, lest they should be destroyed by the rebels; yet the officers of the other corps wished to have them sent out of the town, and to this effect joined in a memorial to government. Some of these officers disliked my father, from differences of electioneering interests; others, from his not having kept up an acquaintance with them; and others, not knowing him in the least, were misled by party reports and misrepresentations.

These petty dissensions were, however, at one moment suspended and forgotten in a general sense of danger. An express arrived late one night, with the news that the French, who were rapidly advancing, were within a few miles of the town of Longford. A panic seized the people. There were in the town eighty of the carabineers and two corps of yeomanry, but it was proposed to evacuate the garrison. My father strongly opposed this measure, and undertook, with fifty men, if arms and ammunition were supplied, to defend the gaol of Longford, where there was a strong pass, at which the enemy might be stopped. He urged, that a stand might be made there, till the king's army should come up. The offer was gladly accepted—men, arms, ammunition, all he could want or desire, were placed at his disposal. He slept that night in the gaol, with every thing prepared for its defence; but the next morning fresh news came, that the French had turned off from the Longford road, and were going towards Granard; of this, however, there was no certainty. My father, by the desire of the commanding officer, rode out to reconnoitre, and my brother went to the top of the court-house with a telescope for the same purpose. We (Mrs. Edgeworth, my aunts, my sisters, and myself) were waiting to hear the result in one of the upper sitting-rooms of the inn, which fronted the street. We heard a loud shout, and going to the window, we saw the people throwing up their hats, and heard huzzas. An express had arrived with news, that the French and the rebels had been beaten; that General Lake had come up with them at a place called Ballynamuck, near Granard; that 1,500 rebels and French were killed, and that the French generals and officers were prisoners.

We were impatient for my father, when we heard this joyful news; he had not yet returned, and we looked out of the windows in hopes of seeing him, but we could see only a great number of the people of the town, shaking hands with each other. This lasted a few minutes, and then the crowd gathered in silence round one man, who spoke with angry vehemence and gesticulation; stamping, and frequently wiping his forehead. We thought he was a mountebank haranguing the populace, till we saw, that he wore a uniform. Listening with curiosity to hear what he was saying, we observed, that he looked up towards us, and we thought we heard him pronounce the names of my father and brother in tones of insult. We could scarcely believe what we heard him say. Pointing up to the top of the court-house, he exclaimed, "*That* young Edgeworth ought to be dragged down from the top of that house."

Our housekeeper burst into the room, so much terrified she could hardly speak.

"My master, ma'am!—it is all against my master, the mob say they will tear him to pieces, if they catch hold of him. They say he's a traitor, that he illuminated the gaol to deliver it up to the French."

No words can give an idea of our astonishment. Illuminated! what could be meant by the gaol being illuminated? My father had literally but two farthing candles, by the light of which he had been reading the newspaper late the preceding night. These however were said to be signals for the enemy! The absurdity of the whole was so glaring, that we could scarcely conceive the danger to be real; but our pale landlady's fears were urgent, she dreaded that her house should be pulled down. We found, that the danger was not the less because the accusation was false; on the contrary, it was great in proportion to its absurdity, for the people who could at once be under such a perversion of intellects, and such an illusion of their senses, must indeed be in a state of frenzy.

The crowd had by this time removed from before the windows; but we heard, that they were gone to that end of the town, through which they expected Mr. Edgeworth to return.

We wrote immediately to the commanding officer, informing him of what we had heard, and requesting his advice and assistance. He came to us, and recommended, that we should send a messenger to warn Mr. E. of his danger, and to request, that he would not return to Longford this day. The officer added, that, in consequence of the rejoicings for the victory, his men would probably be all drunk in a few hours, and that he could not answer for them. This officer, a captain of yeomanry, was a good-natured, but inefficient man, who spoke under considerable nervous agitation, and seemed desirous to do all he could, but not to be able to do any thing. We wrote instantly, and with difficulty found a man, who undertook to convey the note. It was to be carried to meet him on one road, and Mrs. Edgeworth and I determined to drive out to meet him on the other. We made our way down a back staircase into the inn yard, where the carriage was ready. Several gentlemen spoke to us as we got into the carriage, begging us not to be alarmed: Mrs. Edgeworth answered, that she was more surprised than alarmed. The commanding officer and the sovereign of Longford walked by the side of the carriage through the town; and as the mob believed, that we were going away not to return, we got through without molestation. We went a few miles on the road towards Edgeworth town, till at a tenant's house we heard, that my father had passed by half an hour ago; that he was riding in company with an officer, supposed to be of Lord Cornwallis's or General Lake's army; that they had taken a *short cut,* which led into Longford by another entrance. Most fortunately not that at which an *armed* mob had assembled, expecting the object of their fury. Seeing him return to the inn with an officer of the king's army, they imagined, as we were afterwards told, that he was brought back a prisoner, and they were satisfied.

The moment we saw him safe, we laughed at our own fears, and again doubted the reality of the danger, more especially as he treated the idea with the utmost incredulity and scorn.

Major (now General) Eustace was the officer, who returned with him. He dined with us; every thing appeared quiet. The persons who had taken refuge at the inn were now gone to their homes, and it was supposed, that, whatever dispositions to riot had existed, the news of the approach of some of Lord Cornwallis's suite, or of troops who were to bring in the French prisoners, would prevent all probability of disturbance. In the evening the prisoners arrived at the inn; a crowd followed them, but quietly. A sun-burnt, coarse looking man, in a huge cocked hat, with a quantity of gold lace on his clothes, seemed to fix all attention; he was pointed out as the French General Homberg, or Sarrazin. As he dismounted from his horse, he threw the bridle over its neck, and looked at the animal as being his only friend.

We heard my father in the evening ask Major Eustace, to walk with him through the town to the barrack-yard to evening parade; and we saw them go out together without our feeling the slightest apprehension. We remained at the inn. By this time Col. Handfield, Major Cannon, and some other officers, had arrived, and they were at the inn at dinner in a parlour on the ground-floor, under our room. It being hot weather, the windows were open. Nothing now seemed to be thought of but rejoicings for the victory. Candles were preparing for the illumination; waiters, chambermaids, landlady, all hands were busy scooping turnips and potatoes for candlesticks, to stand in every pane of every loyal window.

In the midst of this preparation about half an hour after my father had left us, we heard a great uproar in the street. At first we thought the shouts were only rejoicings for victory, but as they came nearer we heard screechings and yellings, indescribably horrible. A mob had gathered at the gates of the barrack yard, and joined by many soldiers of the yeomanry on leaving parade, had followed Major Eustace and my father from the barracks. The major being this evening in colored clothes, the people no longer knew him to be an officer; nor conceived, as they had done before, that Mr. Edgeworth was his prisoner. The mob had not contented themselves with the horrid yells that we had heard, but had been pelting them with hard turf, stones, and brickbats. From one of these my father received a blow on the side of his head, which came with such force as to stagger, and almost to stun him; but he kept himself from falling, knowing that if once he fell he should be trampled under foot. He walked on steadily till he came within a few yards of the inn, when one of the mob seized hold of Major Eustace by the collar. My father seeing the windows of the inn open, called with a loud voice, "Major Eustace is in danger!"

The officers, who were at dinner, and who till that moment had supposed the noise in the street to be only drunken rejoicings, immediately ran out, and rescued Major Eustace and my father. At the sight of British officers and drawn swords, the populace gave way, and dispersed in different directions.

The preparation for the illuminations then went on, as if nothing had intervened. All the panes of our windows in the front room were in a blaze of light by the time the mob returned through the street. The night passed without further disturbance.

As early as we could the next morning we left Longford, and returned homewards, all danger from rebels being now over, and the rebellion having been terminated by the late battle.

When we came near Edgeworth-Town, we saw many well known faces at the cabin doors, looking out to welcome us. One man, who was digging in his field by the road side, when he looked up as our horses passed, and saw my father, let fall his spade and clasped his hands; his face, as the morning sun shone upon it, was the strongest picture of joy I ever saw. The village was a melancholy spectacle; windows shattered, and doors broken. But though the mischief done was great, there had been little pillage. Within our gates we found all property safe; literally "not a twig touched, nor a leaf harmed." Within the house every thing was as we had left it—a map that we had been consulting was still open on the library table, with pencils, and slips of paper containing the first lessons in arithmetic, in which some of the young people had been engaged the morning we had been driven from home; a pansy, in a glass of water, which one of the children had been copying, was still on the chimney-piece. These trivial circumstances, marking repose and tranquillity, struck us at this moment with an unreasonable sort of surprise, and all that had passed seemed like an incoherent dream. The joy of having my father in safety remained, and gratitude to Heaven for his preservation. These feelings spread inexpressible pleasure over what seemed to be a new sense of existence. Even the most common things appeared delightful; the green lawn, the still groves, the birds singing, the fresh air, all external nature, and all the goods and conveniencies of life, seemed to have wonderfully increased in value, from the fear into which we had been put of losing them irrecoverably.

The first thing my father did, the day we came home, was to draw up a memorial to the Lord Lieutenant, desiring to have a court martial held on the sergeant, who, by haranguing the populace, had raised the mob at Longford; his next care was to walk through the village, to examine what damage had been done by the rebels, and to order that repairs of all his tenants' houses should be made at his expense. A few days after our return, Government ordered, that the arms of the Edgeworth-Town infantry should be forwarded by the commanding officer at Longford. Through the whole of their hard week's trial, the corps had, without any exception, behaved perfectly well. It was perhaps more difficult to honest and brave men passively to bear such a trial, than any to which they could have been exposed in action.

When the arms for the corps arrived, my father, in delivering them to the men, thanked them publicly for their conduct, assuring them, that he would remember it, whenever he should have opportunities of serving them, collectively or individually. In long after years, as occasions arose, each, who continued to deserve it, found in him a friend, and felt, that he more than fulfilled his promise. Now that he could look back upon suspicions and accusations, now that events had decided upon guilt and innocence, and had shown who were, and who were not, implicated in this rebellion, he had the satisfaction of feeling, that, though he had trusted much, he had not in *many* instances, trusted rashly. In some few cases he was deceived.—Who in Ireland at that time can boast, that he was not? Some few, very few indeed, of his tenantry, on a remote estate—alas too near Ballynamuck, did join the rebels! These persons were never readmitted on my father's estate.

In some cases it was difficult to know what ought to be done: for instance, with regard to the man who had saved his house from pillage, but who had joined the rebels. It was the wise policy of Government, to pardon those, who had not been ringleaders in this rebellion; and who, repenting of their folly, were desirous to return to their allegiance, and to their peaceable duties. My father sent for this man, and said he would apply to Government for a pardon for him. The man smiled, and clapping his pocket said, "I have my *Corny* here safe already, I thank your Honor—else sure I would not have been such a fool, as to be shewing myself without I had a *purtection*."—A pardon, signed by the Lord Lieutenant, Lord Cornwallis, in their witty spirit of abbreviation, they called a *Corny*.

We observed, and thought it an instance of Irish acuteness and knowledge of character, that this man was sure my father never would *forget him*, though he gave him nothing at this time. When my father said, that, though we were obliged to him for saving the house, he could not *reward* him for being a rebel; he answered, "Oh, I know that I could not expect it, nor look for any thing at all, but what I got—*thanks*." With these words he went away satisfied.

A considerable time afterward my father, finding that the man conducted himself well, took an opportunity of serving him. Rewards my father thought fully as necessary and as efficacious as punishments, for the good government of human creatures: therefore he took especial care, not only to punish those who had done ill, but to reward, as far as he could, all who had done well.

Before we quit this subject, it may be useful to record, that the French generals, who headed this invasion, declared they had been completely deceived as to the state of Ireland. They had expected to find the people in open rebellion, or, at least, in their own phrase, *organized* for insurrection; but to their dismay, they found only ragamuffins, as they called them, who, in joining their standard, did them infinitely more harm than good. It is a pity, that the lower Irish could not hear the contemptuous manner, in which the French, both officers and soldiers, spoke of them and of their country. The generals described the stratagems, which had been practised upon them by their good allies. The same rebels frequently returning with different tones and new stories, to obtain double and treble provisions of arms, ammunition, and uniforms—selling the ammunition for whiskey, and running away at the first fire in the day of battle. The French, detesting and despising those by whom they had been thus cheated, pillaged, and deserted, called them beggars, rascals, and savages. They cursed also without scruple their own Directory, for sending them, after they had, as they boasted, conquered the world, to be at last beaten in an Irish bog. Officers and soldiers joined in swearing, that they would never return to a country, where they could find neither bread, wine, nor discipline; and where the people lived on roots, whiskey, and lying.

These, as my father observed, were comfortable words for Ireland. National antipathy, thus felt and expressed between the invaders and the Irish malecontents, would be better security for the future quiet and safety of the country even than the victory. Full of these thoughts, unconcerned about himself, and in excellent spirits, he succeeded in turning our attention to new objects. The Longford mob completely vanished from our imagination. Reflecting upon what had passed, my father drew from it one useful conclusion for his own future conduct—that he ought to mix more with society, and make himself more generally known in Ireland.

At all times it is disadvantageous to those, who have the reputation of being men of superior abilities, to seclude themselves from the world. It raises a belief, that they despise those with whom they do not associate; and this supposed contempt creates real aversion. The being accused of pride or singularity may not, perhaps, in the estimation of some lofty spirits and independent characters, appear too great a price to pay for liberty and leisure; they will care little, if they be misunderstood or misrepresented by the vulgar; they will trust to truth and time, to do them justice. This may be all well in ordinary life, and in peaceable days: but in civil commotions, the best and the wisest, if he have not made himself publicly known, so as to connect himself with the interests and feelings of his neighbours, will find none to answer for his character, if it be attacked, or to warn him of the secret machinations of his enemies; none who on any sudden emergency will risk their own safety in his defence: he may fall and be trampled upon by numbers, simply because it is nobody's business or pleasure, to rally to his aid. Time and reason may right his character, and may bring all who have injured, or all who have mistaken him, to repentance and shame, but in the interval he *must* suffer—he *may* perish. There is no absurdity of ignorance, or grossness of calumny, from which he may hope to be secure. He may be conceived to be a traitor, because he would not be a tyrant; he may be called a rebel, for offering to defend a loyal garrison; and may well nigh be torn to pieces by a mob, for having read the newspapers by two farthing candles.

CHAPTER XI.

GOVERNMENT, having at this time the union between Great Britain and Ireland in contemplation, were desirous, that the Irish aristocracy and country gentlemen should be convinced of the kingdom's insufficiency to her own defence, either against invasion or internal insurrection; with this view it was politic, to let the different parties struggle with each other, till they completely felt their weakness and their danger. It was said, with what justice I do not presume to determine, that with this view the *Orange-men* and *United Irishmen*, the *Ribbon-men*, the *Thrashers*, the *Croppies*, the *Caravats*, and *Shanavats*, or by whatever other names, more savage or more ridiculous, they called themselves, were let loose upon each other. It is certain, that the combinations of the disaffected at home, and the advance of foreign invaders, were not checked till the peril became imminent, and till the purpose of creating universal alarm had been fully effected. As soon as the Commander in Chief and the Lord Lieutenant (at that time joined in the same person) exerted his full military and civil power, the invaders were defeated, and the rebellion was extinguished. The nightly depredations, and the nocturnal *scouring of the country* ceased; the dogs of war were recalled, and the havoc ended. Those would-be-soldiers, who, under the cry of loyalty, had become hunters of men, found their occupation gone. In the uniform of a hunt, instead of the uniform of a military corps, they were obliged to recur to their former sports, and to content themselves with pursuing no "other game than beast or bird." The petty magisterial tyrants, who had been worse than vain of their little brief authority, were put down, or rather, being no longer upheld, sunk to their original and natural insignificance. The laws returned to their due course; and, with justice, security and tranquillity were restored.

My father honestly, not ostentatiously, used his utmost endeavours, to obliterate all that could tend to perpetuate ill-will in the country. Among the lower classes in his neighbourhood he endeavoured to discourage that spirit of recrimination and retaliation, which the lower Irish are too prone to cherish, and of which they are proud. "*Revenge is sweet, and I'll have it,*" were words, which an old beggar-woman was overheard muttering to herself as she tottered along the road.

With this pervading sentiment the poor consider legal revenge as one of the sweetest luxuries, which wealth puts in the power of their superiors in station. Assuredly my father did some service in his neighbourhood, by shewing, that it was not among the number of his luxuries, and by discountenancing the fashion for this perverted taste.

It sometimes happens, that persons of the higher ranks preach in words that abstinence from resentment, that spirit of forgiveness, which they do not practise. Moral advice in this case has no good effect.

The lower Irish are such acute observers, that there is no deceiving them as to the state of the real feelings of their superiors. They know the signs of what passes within, more perfectly than any physiognomist, who ever studied the human face, or human head. Combining quickly every circumstance in the manners, gestures, and slightest actions of those, whom it is their interest to study; the result is, that the ruling, or the reigning passion can scarcely escape their detection.

It was soon seen by all of those, who had any connexion with him, that my father was sincere in his disdain of vengeance,—of this they had convincing proof in his refusing to listen to the tales of slander, which so many were ready to pour into his ear, against those who had appeared to be his enemies.

They saw that he determined to have a public trial of the man who had instigated the Longford mob, but that for the sake of justice, and to record what his own conduct had been; he did not seek this trial from any paltry motives of personal resentment.

His thoughts were soon called to that most important question, of the union between England and Ireland, which it was expected would be discussed at the meeting of parliament.

It was late in life to begin a political career, imprudently so, had it been with the common views of family advancement, or of personal fame; but his chief hope in going into parliament was to obtain assistance in forwarding the great object of improving the education of the people; he wished also to assist in the discussion of the union. He was not without a natural desire, which he candidly avowed, to satisfy himself how far he could succeed as a parliamentary speaker, and how far his mind would stand the trial of political competition, or the temptations of ambition.

On the subject of the union he had not yet

been able, in parliamentary phrase, *to make up his mind;* and he went to the house in that state, in which so many profess to find themselves, and so few ever really are, anxious to hear the arguments on both sides, and open to be decided by whoever could shew him that, which was best for his country.

The debate on the first proposal of the union was protracted to an unusual length, and when he rose to speak, it was late at night, or rather it was early in the morning, two o'clock—the house had been so wearied, that many of the members were asleep. It was an inauspicious moment. No person present, not even the speaker, who was his intimate friend, could tell on which side he would vote.—Curiosity was excited: some of the outstretched members were rouzed by their neighbours, whose anxiety to know on which side he would vote prompted them to encourage him to proceed. This curiosity was kept alive as he went on; and when people perceived, that it was not a *set* speech, they became interested. He stated his doubts just as they had really occurred, balancing the arguments as he threw them by turns into each scale, as they had balanced one another in his judgment; so that the doubtful beam nodded from side to side, while all watched to see when its vibrations would settle. All the time he kept both parties in good humour, because each expected to have him their own at last. After stating many arguments in favor of what appeared to him to be the advantages of the union; he gave his vote against it, because he said he had been convinced by what he had heard in that house this night, that the union was at this time decidedly against the wishes of the great majority of men of sense and property in the nation. He added, that, if he should be convinced that the opinion of the country changed at the final discussion of the question, his vote would be in its favor.

One of the anti-unionists, who happened not to know my father personally, imagined from his accent, style, and manner of speaking, that he was an Englishman; and accused Government of having brought a new member over from England, to impose him upon the house as an impartial country gentleman, who was to make a pretence of liberality by giving a vote against the union, while by arguing in its favor, he was to make converts for the measure. Many on the ministerial bench, who had still hopes, that on a future occasion Mr. E. might be *convinced*, and brought to vote with them, complimented him highly, declaring that they were completely surprised, when they learnt how he voted; for that undoubtedly the best arguments on their side of the question had been produced in his speech. Lord Castlereagh found the measure so much against the sense of the house, that he pressed it no further at that time.

This session my father had the satisfaction of turning the attention of the house to a subject, which he considered to be of greater and more permanent importance than the union, or than any merely political measure could prove to his country, *the education of the people.* By his exertions, a select committee was appointed, and they adopted the resolutions drawn up by him. When the report of this committee was brought up to the house, my father spoke at large upon the subject. As he really spoke extempore, as he never prepared sentences, kept any note, or had any power of repeating in the same words what he had once uttered, he could not afford any assistance to the reporter, and what he said was imperfectly given in the public papers. I do not attempt however any alteration or correction; all who are used to writing, or to public speaking, will feel, that it would be easy to have dressed it up; but I prefer giving it in the unaltered newspaper report.

Monday, Feb. 25, 1799.

"The house having resolved into a committee, Mr. Röchford in the chair, on the report of the select committee appointed to take into consideration the best means of educating the children of the lower classes—

" Mr. *Edgeworth* rose, and called the attention of gentlemen to a subject of the highest importance to the peace and prosperity of the country, he meant the education of the peasantry and working part of the community. He had talked over this subject with many gentlemen, and he was bold to say, that if the people were better instructed, they would be better subjects.

" It was not his intention to submit any deviation from the leng established plan of education, such as was adopted by the commissioners appointed in the year 87 for that purpose; but he felt anxious to remove prejudices, if any prejudices remained in the minds of gentlemen against the education of the lower classes. On this subject there was vast room for speculation; he did not signify by the word speculation, any new wild theory; he meant to proceed on matter of fact, on the deduction of facts accurately ascertained; this was the right difference between theory and practice, and in it consisted that true useful speculation which was founded in experience. It was by collecting a number of facts in any case, and by drawing deductions from them, that mankind were properly directed in scientific and moral pursuits. He was sensible that some were averse, not he hoped in that House, to improving the minds of the working people; but such opinions were grounded in error, and a false estimate of human nature. It would appear that in countries where the people at large were best instructed, that they were best affected to constituted authority and regular Government. Such was the situation of Scotland, where though the people in general have more knowledge and learning than in any other country, the pernicious principles which destroyed other parts of Europe could not be propagated.

" It was impossible, Mr. Edgeworth observed, when moral principles are instilled into the human mind, when people are regularly taught their duty to God and Man, that abominable tenets can prevail to the subversion of subordination and society. He would venture to assert, though the power of the sword was great, that the force of education was greater. It was notorious, that the writings of one man, Mr. Burke, had changed the opinions of the whole people of England against the French revolution; he was not prepared to say, that Mr. Burke's indignation did not carry him too far; but he was sure his book had brought Englishmen to a right understanding on that great subject. Were the common people instructed, Mr. Paine's Rights of Man could not have such prevalence as it obtained; for if we estimated right according to the ideas of Puffendorf, and other writers on the law of nations, it meant something done according to rule. Now the Rights of Man, according to Mr. Paine, were prior and antecedent to all rule; and therefore such revolutionary writings contained nothing to convince any reasonable man. If proper books were circulated through the country, and if the public mind was prepared for the reception of their doctrines, it would be impossible to make the ignorance of the people an instrument of national ruin.

The Irish nation, he knew by experience, have as quick apprehensions, and as good hearts as any people in the world--- he never met men so easily taught, so intelligent when taught, or so calculated to make virtuous, loyal subjects. When they received a benefit, they were disposed to return it; but when their minds were perverted, there existed in them a headstrong principle fatal to themselves, and dangerous to the state. There is, he contended, a fund of goodness in the Irish as well as in the English nature. Did God give different minds to different countries? No! the difference of mind arose from education. It therefore became the duty of Parliament to improve as much as possible the public understanding—for the misfortunes of

Ireland were owing not to the heart, but the head; and the defect was not from nature, but from want of culture.

"He did not wish for the people any other education, but what might afford them a knowledge of their duty, what would make them virtuous and loyal, useful to themselves and to the state. He was sensible that such a change must be owing to gradual operation; at present the strong arm of power should be held over the ignorant; but when down they were not to be oppressed, they should be raised up to the rank of men, by informing their understanding. No man was more ready to give supplies to Government, and power to the Executive—his life, his fortune, and every thing dear to him was on that side; but if the house meant to fix the foundations of private and public security, they must resort to the education of the people.

"By a wise system of education the present bigotry might be rooted out; and our lower classes might have that useful information which would improve the rational mind and secure the state. Though England had many advantages over us, we are, he observed, without one great disadvantage which prevails with them. He meant the consequence of their poor laws, which holds out a premium for vice and intemperance.

"He would move, 1. That it is the opinion of the House, that the state of public education in this country is highly defective, and requires the interposition of Parliament.

"2. That it is the opinion of the House, that one or more schools should be established in each parish.

"3. That it is the opinion of the House, that the masters for such schools should undergo examinations, obtain certificates of their conduct, &c., and be licensed annually by the diocesan.

"4. That it is the opinion of the House, that one or more visitors be appointed to inspect the schools in each parish one or more times in each year."

Leave was granted to bring in a bill for the

improvement of the education of the people of Ireland; and thence proceeded, after the Irish Parliament was no more, the appointment of a Board and Commissioners of Education.

During this session my father spoke again two or three times, on some questions of revenue regulation and excise laws; of little consequence separately considered, but of importance in one respect, in their effect on the morality of the people. He pointed out, that nothing could with more certainty tend to increase the crime of perjury than the multiplying customhouse oaths, and what are termed oaths of office. These are frequently considered as mere forms, and as such are taken and evaded by *gentlemen*, to whose example the lower people look up for the fashion of their morals. Customhouse oaths are frequently both administered and laughed at by persons just a step above the lower classes, whose example operates daily and hourly; and, combined with the view of petty gain and constant impunity, hold forth what may be called a perpetual bounty on perjury. In Ireland the habits of the common people are already too lax with regard to truth. The differences of religion, and the facilities of absolution, present difficulties so formidable to their moral improvement, as to require all the counteracting powers of education, example, public opinion, and law. The legislature should therefore be peculiarly careful, not to increase the evil by fiscal regulations, even were we certain of obtaining by them the partial good they promise: but, in fact, even the hope of this good is illusive. Multiplying oaths injures the revenue, by increasing incalculably the means of evading the very laws and penalties, by which it is attempted to bind the subject. Experience proves, that this is a danger of no small account to the revenue; though trifling when compared with the importance of the general effect on national morality, and on the safety and tranquillity of the state, all which must ultimately rest at all times and in all countries upon religious sanction. "It was not," my father observed, "by increasing pains and penalties, or by any severity of punishment, that the observance of laws can be secured; on the contrary, small but certain punishments, and few but punctually executed laws, are most likely to secure obedience, and to effect public prosperity."

It is with no small satisfaction that I have lately heard these sentiments and opinions of my father's expressed by the best and wisest in the British Senate; enforced and adorned by eloquence the most brilliant, persuasive, and convincing; supported by facts and evidence on the largest scale, which time and numbers, commercial records, and the history of nations can command.

The following letters to Dr. Darwin will shew the effect produced by a political campaign on my father's private feelings:

"*March* 12, 1799.

"MY DEAR DOCTOR,

"ONE morning in the year 1791, when you were confined to your bed, and asked me to come into your room at Derby to take leave of you, you told me, that if ever I went into Parliament I should be torn to pieces by the teeth and fangs of the parliamentary leaders; but in truth, my dear Doctor, I have found them very harmless creatures as to myself individually, though I acknowledge, that some of them can tear their country to pieces with all possible ferocity.

"I then told you I was not ambitious, and I continue to be of the same opinion. For though I have not been entirely unsuccessful in gaining the attention of the House, I do not feel one jot the more disposed to become a commissioner of the revenue, a sinecure placeman, or a pensioner.

"In the shock of contending parties, seizing a moment of temporary calm, I found an opportunity of laying the foundation of a system, by which the wretched poor of this country may in time be rendered less savage. * * *

"God bless you, my dear Doctor, and keep all evil from you, from your good and amiable lady, and from your *giant brood.*

"Yours, affectionately,

"R. L. E."

TO THE SAME.

" *March* 31, 1800.

" MY DEAR AND EXCELLENT FRIEND,

" Words which are not meant as a mere translation of '*vir excellentissime*,' but which really express my feelings and opinions. I have been absent from home upon the business of the nation. The fatigue of the session was enormous. I am a unionist, but I vote and speak against the union now proposed to us—as to my reasons, are they not published in the reports of our debates? &c.

" It is intended to force this measure down the throats of the Irish, though five sixths of the nation are against it. Now, though I think such a union as would identify the nations, so as that Ireland should be as Yorkshire to Great Britain, would be an excellent thing; yet I also think, that the good people of Ireland ought to be *persuaded* of this truth, and not be dragooned into submission.

" The minister avows, that seventy-two boroughs are to be compensated—*i. e.* bought by the people of Ireland with one million and a half of their own money; and he makes this legal by a small majority, made up chiefly of these very borough members. When thirty-eight country members out of sixty-four are against the measure, and *twenty-eight* counties out of *thirty-two* have petitioned against it, this is such abominable corruption, that it makes our parliamentary sanction worse than ridiculous.

" I had the honour of offering, for myself and for a large number of other gentlemen, that, if the minister could by any means win the nation to the measure, and shew us even a small preponderance in his favour, we would vote with him.

" So far for politics. I had a charming opportunity of advancing myself and my family, but I did not think it wise *to quarrel with myself*, and lose my own good opinion at my time of life. What *did* lie in my way for my vote I will not say, but I stated in my place in the House, that I had been offered 3000 guineas for my seat during the few remaining weeks of the session.

" Maria continues writing for children, under the persuasion that she cannot be employed more serviceably.

" I shall now resume my pen, with the same sincere intention of being useful, and I shall write and draw Mechanoiconomia, or a Description of Machinery for Domestic Purposes.

" I have not yet received the whole of Phytologia. Not one copy has yet made its appearance in Ireland.

" I am just going, at the desire of a national institution, similar to your Board of Agriculture, to begin an Agricultural, Economic, and Statistic Account of this County, in which I shall make free with your information.—By the by, I used your name in two debates this session with effect.

" Yours,

" R. L. E."

To the view which his letters exhibit of his sentiments freshly given to his most intimate friend, I add the following pages, written by my father many years afterwards :—(in 1817.)

" I was in the Irish Parliament, when the question of a legislative union between Great Britain and Ireland was finally discussed. The part which I took may be known, by the speeches which I then published. One of these is miserably reported, appearing to be faulty both in sense and grammar; they are, however, sufficient to enable my children to decide upon the consistency of my conduct, which has, I hope, never deviated from what appeared to me to be right and honourable. The *influence* of the Crown was never so strongly exerted as upon this occasion. It is but justice, however, to Lord Cornwallis and Lord Castlereagh, to give it as my opinion, that they *began* this measure with sanguine hopes, that they could convince the reasonable part of the community, that a cordial union between the two countries would essentially advance the interests of both. When, however, the ministry found themselves in a minority, and that a spirit of general opposition was rising in the country, a member of the House, who had been long practised in parliamentary intrigues, had the audacity to tell Lord Castlereagh from his place, that, ' if he did not employ the *usual means of persuasion* on the members of the House, he would fail in his attempt, and that the sooner he set about it the better.'

" This advice was followed; and it is well known, what benches were filled with the proselytes, that had been made by *the convincing arguments*, which obtained a majority. In one of the last sittings of the last session of the last Parliament of Ireland, a distinguished member said he would not sit in that House, to hear its death-warrant read from the chair. It was expected, that he would go out of the House at the close of this speech. When he sat down*, I rose from my seat, went to the bar, and bowing to the Chair, withdrew; I was followed by Mr. George Ponsonby, and by some other members, whose names I do not now remember."

I have been assured, that my father's speeches in Parliament, especially the last, made a considerable impression on the House; and I know, that temptations were held out to him in every possible form, in which they could *flatter* personal ambition or family interest; he had offers of all that could serve or oblige his dearest friends, and choice of situations, in which he might, as it was said, gratify his peculiar tastes, serve his country, and accomplish his favourite object of im-

---

* *Note by the Editor.*

I shewed these pages, Sept. 1817, to Mr. Foster, (the Speaker of the Irish House of Commons); and I asked whether he remembered this circumstance: he assured me, that it was perfectly in his recollection.

M. E.

proving the education of the people. Opportunity for *convenient distinctions* in his case appeared also; since he was avowedly of opinion, that the measure in question would be ultimately advantageous to this country, though he thought the means of carrying it, and the forcing it, at this time, contrary to the sense of the people, was wrong. But, however plausible, he would not admit of any such nice casuistry in a case of conscience; he would not palter with the fiend Ambition *.

---

* Among those of my father's friends, who were most averse to the Union, was Lord Charlemont.—He did not live to see it completed; he died in August, 1799. My father was much attached to him, esteeming him for his public conduct, and loving him for his amiable disposition and polished manners. In Lord Charlemont, as he observed, a desire to please and conciliate were so joined with sincerity, and integrity of character, that he formed an exception to the general rule, that no man can be good for much, who is liked by all parties.

I found among my father's papers the following memorandum. "A few days after the death of Lord Charlemont some English friends of mine, who came to Dublin, were very desirous to see his statues and library; at such a moment it was impossible to apply to any of the family for admission, but the old porter, who knew how intimate I had been with his master, ventured to admit us silently into the library; where I was not a little moved at seeing the chair, on which he usually sat, in it's accustomed place; his gloves and snuff-box on the table; and Practical Education, which he had been reading, lying open upon his chair."

CHAPTER XII.

WHEN it was known in the county and the town of Longford, that Mr. Edgeworth had voted against the union, he became suddenly the idol of those, who but a few months before would have willingly stoned him to death.

At the Spring Assizes, the trial of the man, who had harangued and instigated the mob at Longford, came on. When my father had applied for a court martial, some difficulties had occurred.—It was found, that, though this man was a permanent sergeant of yeomanry, yet, as he had not subscribed to the articles of war, he could not be tried by a court martial, and he was therefore prosecuted in a civil suit. During the course of the trial, it appeared, that this sergeant was a mere ignorant enthusiast, who had been worked up to phrenzy by some, more designing than himself. Having accomplished his own object of publicly proving every fact that concerned his own honor and character, my father felt desirous, that the poor culprit, who was now ashamed and penitent, should not be punished. The evidence was not pressed against him, and he was acquitted. My father's counsel was his zealous and eminent friend, Mr. (afterward Judge) Fox. The judge, who presided at the trial, was Sir Michael Smith, whose charge, as I have been assured, was able and eloquent; as honorable to himself, as it was to my father. As they were leaving the Court House, my father saw, and spoke in a playful tone, to the penitent sergeant, who, among his other weaknesses, happened to be much afraid of ghosts.—"Sergeant, I congratulate you," said he, "upon my being alive here before you—I believe you would rather meet me than my ghost!"—Then cheering up the man with the assurance of his perfect forgiveness, he passed on.

The malevolent passions my father always considered as the greatest foes to human felicity—they would not stay in his mind—he was of too good and too happy a nature. He forgot all, but the moral which he drew for his private use from this Longford business—He kept ever afterwards the resolution he had made, to mix more with general society.

While he was in Parliament, he spent two winters in Dublin, and went in the Spring of 1799, to England. He visited his old friends, Mr. Keir, Mr. Watt, Dr. Darwin, and Mr. William Strutt, of Derby. In passing through different parts of the country, he saw, and delighted in shewing us, every thing curious and interesting in art and nature. Travelling, he used to say, was from time to time necessary, to change the course of ideas, and to prevent the growth of local prejudices.

He went to London, and paid his respects to his friend Sir Joseph Banks, attended the meetings of the Royal Society, and met various old acquaintance, whom he had formerly known abroad;—some of them persons of rank and fortune, who, since he had seen them, had much improved from the wildness of youth, and now appeared with mature and sedate characters. He saw also others, whom he had formerly known on the continent, some who had been the tutors or travelling companions of these young noblemen; and who having since risen by literature to fame and affluence, were equal, at least in consideration, to those, who, by birth, had been so much their superiors. All these persons recognized my father with cordial pleasure, recollecting every circumstance of his works at Lyons, of his conduct there, and of the high estimation in which he had been held. These testimonies were peculiarly agreeable coming home to his children and family. Accidental recurrence of the recollection of the friends of a man's youth, and the witnesses of what he has done, and has been, at different periods, should be noticed in every life, as a source of considerable pain or pleasure, self-complacency or self-reproach.

My father's having become an author put it in his power, during this journey, to enlarge his acquaintance in the English literary world. Many men of learning, whom he had formerly known, recognized him with satisfaction, and introduced him to several contemporary authors. Among the friends he formed during this summer in England, and in consequence of the publication of his sentiments on education, was Mrs. Barbauld. Her writings he had long admired for their classical strength and elegance, for their high and true tone of moral and religious feeling, and for their practically useful tendency. Ever the true friend and champion of female literature, and zealous for the honour of the female sex, he rejoiced with all the enthusiasm of a warm heart, when he found, as he now did, female genius guided by feminine discretion. He exulted in every instance of literary celebrity, supported by the amiable and respectable virtues of private life; proving by example, that the cultivation of female talents does not unfit women for their domestic duties and situation in society. After a short acquaintance formed in London with Mrs. Barbauld, she gratified him by accepting an invitation, to pass some time with us at Clifton; and ever afterwards, though thrown at a great distance from each other, her constant friendship for him was a source of great pleasure and just pride.

In his own account of his earlier life he has never failed to mark the time and manner of the commencement of valuable friendships, with the same care and vividness of recollection, with which some men mark the date of their obtaining promotion, places, or titles. I follow the example he has set me.

While we were at Clifton, when Mrs. Edgeworth was recovering after the birth of her first child, her two brothers, the Rev. William Beaufort, and Captain Beaufort, of the Royal Navy (whom my father had never seen before), came to visit her. It seldom happens, that, when two large families are connected by marriage, they suit each other in every branch of the connexion. My father's and Mrs. Edgeworth's families were both numerous, and among such numbers, even granting the dispositions to be excellent, and the understandings cultivated, the chances were against their suiting; but, happily, all

* Author of "Karamania—a description of the South coast of Asia Minor."

the individuals of the two families, though of various talents, ages, and characters, did from their first acquaintance, coalesce. Beyond the most sanguine hopes, that could have been formed, my father was blessed with a new race of friends in this family, springing up to supply to him in age the loss of the distinguished and beloved companions of his youth. After he had lost such a friend as Mr. Day, so tried, so faithful, of such superior abilities, of such a generous, noble character, who could have dared to hope, that he should ever have found another equally deserving to possess his whole confidence and affection? yet such a one it pleased God to give him—and to give him in the brother of his wife. And never man felt more strongly grateful for the double blessing. To Captain Beaufort he became as much attached, as he had ever been to Lord Longford, or to Mr. Day.

His father-in-law, Dr. Beaufort, was also particularly agreeable to him as a companion, and useful as a friend. From his amiable, popular character in Ireland, and from his extensive acquaintance, possessing peculiarly the power of serving his friends in society, Dr. Beaufort knew how to contradict the misrepresentations of party, and to rectify the mistakes of ignorance, without ever exposing those he loved to envy, by indiscreet praise. Such a friend was peculiarly advantageous to my father. The being more known in England was, upon reverberation, useful to him in his own country. He returned to Edgeworth-Town, in the Autumn of 1799.

The latter end of the year 1800 was marked with melancholy, by the death of his daughter Elizabeth, the eldest of the daughters of his last wife; another loss to her parents of the promising fruits of careful education. She lived till she was nineteen, and then fell a sacrifice to hereditary consumption. She died at Clifton, August, 1800.

Two years elapsed afterwards, without leaving a trace on my memory of any event worth recording.

In the year 1802, at the time of the short peace, many travellers, who had visited the continent, returned enraptured with Paris. Letters and visits from some of these, excited my father's curiosity and interest; he was charmed by their accounts of the literary and scientific society in France. He wished to see their public institutions, and was very desirous,that Mrs. Edgeworth's taste for painting should be gratified by the sight of the superb gallery of the Louvre, where the masterpieces of the finest artists, torn from their native countries, had been forced into one spot; the spoils of military despotism gracing the triumph of republican liberty.

Among the foreigners, who came to England about this time, was Professor Pictet * of Geneva, brother of the editor of the Journal Britannique, who translated Practical Education, and with whom my father had had some correspondence on the subject. Professor Pictet visited Ireland, and came to Edgeworth-Town. He decided us to go abroad, by the kind offers of introduction to numerous literary friends at Paris; and assurances, that from what they already knew of him, through his writings on Education, they were prepared to receive him and his family with cordiality. The tour was arranged for the ensuing Autumn, and the pleasure of revisiting some of his old English friends, Dr. Darwin in particular, was full in his contemplation, when he received the following letter.

FROM DR. DARWIN TO MR. EDGEWORTH.

" *Priory, near Derby, April* 17, 1802.

" DEAR EDGEWORTH,

" I am glad to find, that you still amuse yourself with mechanism, in spite of the troubles of Ireland.

* Author of " Voyage en Angleterre."

"The *use* of turning aside, or downwards, the claw of a table, I don't see; as it must then be reared against a wall, for it will not stand alone. If the use be for carriage, the feet may shut up, like the usual brass feet of a reflecting telescope.

"We have all been now removed from Derby about a fortnight, to the Priory, and all of us like our change of situation. We have a pleasant house, a good garden, ponds full of fish, and a pleasing valley somewhat like Shenstone's—deep, umbrageous, and with a talkative stream running down it. Our house is near the top of the valley, well screened by hills from the east, and north, and open to the south, where, at four miles distance, we see Derby tower.

"Four or more strong springs rise near the house, and have formed the valley, which, like that of Petrarch, may be called *Val chiusa,* as it begins, or is shut, at the situation of the house. I hope you like the description, and hope farther, that yourself and any part of your family will sometime do us the pleasure of a visit.

"Pray tell the authoress, that the water-nymphs of our valley will be happy to assist her next novel.

"My bookseller, Mr. Johnson, will not begin to print the Temple of Nature, till the price of paper is fixed by Parliament. I suppose the present duty is paid * * * * * *

At these words Dr. Darwin's pen stopped. What follows was written on the opposite side of the paper by another hand.

Sir,

"This family is in the greatest affliction. I am truly grieved to inform you of the death of the invaluable Dr. Darwin. Dr. Darwin got up apparently in health; about eight o'clock, he rang the library bell. The servant, who went, said, he appeared fainting. He revived again,—Mrs. Darwin was immediately called. The Doctor spoke often, but soon appeared fainting; and died about nine o'clock.

"Our dear Mrs. Darwin and family are inconsolable: their affliction is great indeed, there being few such husbands or fathers. He will be most deservedly lamented by all, who had the honor to be known to him.

"I remain, Sir,

"Your obedient humble Servant,

"S. M.

"P. S.—This letter was begun this morning by Dr. Darwin himself."

The shock, which my father felt, must in some degree be experienced by every person, who reads this letter, where the playfulness of the beginning is in such contrast to the end. There is, in the sudden stroke of death, something that no human creature can behold with indifference, even when it falls on one quite unconnected with ourselves, or on one, who had in no way distinguished himself from his fellow mortals; but how much more awfully the blow resounds through the world, when it levels to the dust one preeminent in talent! and how deeply the stroke comes home to the heart, when he, for whose departed genius the public mourns, is a private friend, endeared by all the recollections of generous affection in youth, and of constant friendship through the vicissitudes of life!

As it had happened to my father after the

loss of his friend Mr. Day, he was, after the death of Doctor Darwin, called upon almost immediately to defend his memory against some of those, whose dreadful trade it seems to be, to rob the dead of reputation. He repelled the attack with all the force and indignation of truth. The widow and the sons of Dr. Darwin expressed their gratitude for his affectionate and successful zeal, and joined in the wish, that he would undertake to write the life of his celebrated friend. But difficulties occurred; and in the mean time Miss Seward's memoirs of Darwin, and her critical remarks on his writings, appeared. Of these, and of the general merit of the biographer, and of the poet, the public have necessarily formed their irrevocable judgment; but there occurred one point, which, as it depends much on private opinion and evidence, seemed subject to appeal; and as it affects more the moral character than the literary fame of Darwin, appeared to call for the vindication of friendship. The following letter was addressed to one, whose generous spirit my father loved even more than he admired that genius, which all the world of literature reveres.

TO WALTER SCOTT.

" *Edgeworth-Town, Feb. 3d,* 1812.

DEAR SIR,

"Scales and the sword are emblems as properly suited to the chair of the critic, as to the seat of the municipal justice. I do not therefore hesitate to represent to your court, that an imputation against the fairness of Dr. Darwin's character as an author has been sanctioned by your tribunal, without sufficient evidence to support it.

"It had been said by Miss Seward, that the Doctor had inserted some of her poetry in his Botanic Garden, without making any acknowledgement of his having received it from her; and hence it is concluded, that he meant to pass her verses for his own.

"I was at Lichfield when the lines in question were written by Miss Seward, and I considered them as complimentary to the Doctor, but not as an offering of assistance. The Doctor had not at that time formed the scheme of the Botanic Garden; but many of the lines, which it contains, had been seen by his friends, several years before the garden, which became the theme of his poetry, was in existence. Doctor Darwin composed and wrote the detached pictures in his poem, as he travelled in his carriage among his patients; and these lines were shewn from time to time to his intimate acquaintance, before they were arranged, as they now appear, in one collection. Among these friends I had the honor to be ranked; and for one, I steadily combated his intention of giving to them the form and name, which are before the public. I suggested the scheme of a poetical pantheon, to which the Doctor listened; but he could not prevail upon himself to untwine the amatorial bands, by which he had married the lovesick beings of his vegetable world.

"When I received the printed copy of his work, I expressed my surprise at seeing Miss Seward's lines at the beginning of the poem. He replied, that it was a compliment, which he thought himself bound to pay to the lady, though the verses were not of the same tenor as his own.

" Miss Seward's ode to Captain Cook stands deservedly high in the public opinion. Now, to my certain knowledge, most of the passages, which have been selected in the various reviews of that work, were written by Dr. Darwin. Indeed they bear such strong internal marks of the Doctor's style of composition, that they may easily be distinguished by any reader, who will take the trouble to select them. I remember them distinctly to have been his, and to have read them aloud before Miss Seward and Doctor Darwin, in presence of Sir Brooke Boothby, who will corroborate my assertion.

" I knew the late Dr. Darwin well, and it was as far from his temper and habits, as it was unnecessary to his acquirements, to beg, borrow, or steal, from any person upon earth.

" The indifference of friends to living friends is frequently to be deplored. Fewer still will risk any thing, to support the fame of the deceased. I cannot however refrain from endeavouring, to rescue the moral character of a friend, when it is in my power to give perfect evidence, that it has been misrepresented. If I can by any means rescue Dr. Darwin's name from an unmerited censure, I shall think that I have done for him, what, in a similar situation, I am sure he would have done for me.

" I have the honor to be,

" Dear Sir,

" Your most obedient servant,

" RICH. LOVELL EDGEWORTH."

Perhaps the matter in dispute may not even by this evidence be decided in the minds of some, and it may be thought by others not worth debating further; but at all events, this letter, which was returned to me by the kindness of Walter Scott, is so characteristic of the warmth of my father's heart, that it ought not to be suppressed.

It sometimes happens, that the injudicious zeal of friends, to answer and refute attacks upon the reputation, either of the living or the dead, gives them importance and duration. Many accusations are thus repeated, till they are currently reported; and then the argument for their being believed is, that they have been generally heard. Thus charges, which would never otherwise have been known beyond the limited circle of a man's private friends or foes, are sometimes published to a whole nation, and spread to foreign countries. Slander, thus prolonged beyond the natural and usual period of its ephemeral existence, survives for posterity, and poisons another age. Observing this, some fall into the contrary extreme, and neglect to repel all attacks, and suffer the living and the dead to be calumniated without defence. This is cowardly and base; my father I hope preserved the course most honorable to himself, and advantageous to the memory of his friends.

CHAPTER XIII.

In the autumn of 1802, my father, with Mrs. Edgeworth, my sisters Emmeline and Charlotte, and myself, went to England, where, in October, 1802, my sister Emmeline was married to John King, Esq., and settled at Clifton. Charlotte, who was then, according to the description of a celebrated foreigner, "*jeune personne de seize ans, jolie, fraîche comme la rose, et avec des yeux pleins d'intelligence,*" accompanied us to Paris.

In passing through England we went to Derby, and to the Priory, to which we had been so kindly invited by him who was now no more. The Priory was all stillness, melancholy, and mourning. It was a painful visit, yet not without satisfaction; for my father's affectionate manner seemed to soothe the widow and daughters of his friend, who were deeply sensible of the respect and zealous regard he shewed for Dr. Darwin's memory.

We pursued our journey to France, and, after a delightful tour through the Low Countries, arrived at Paris, where we were to spend the winter. In the Hotel, Place de

Louis Quinze, to which we drove on entering Paris, my father was fortunate in meeting with his illustrious friend, Mr. Watt*. To him he owed an introduction to many foreigners of celebrity. M. Pictet had, as we found, in the most friendly manner prepared the way for us at Paris; and there he more than kept all his promises of assistance, and of introduction to his numerous literary acquaintance, and to a highly cultivated and agreeable society. He was not in Paris on our arrival; but we had, among other kind friends, in particular the venerable Abbé Morellet. During my father's former residence in France, at the time when he was engaged in directing the works of the Rhone and Saone at Lyons, as he mentions in his Memoirs, he wrote a treatise on the construction of mills. He wished that D'Alembert should read it, to verify the mathematical calculations, and for this purpose he had put it into the hands of Morellet. D'Alembert approved of the essay; and my father

* Since this page was written, Mr. Watt has died; and a hand equal to do justice to his merits, has already given to the public a character of him, who among men of invention and science, is justly considered as the most illustrious of the age, and who in domestic life, has been most esteemed and beloved for his amiable temper and unassuming manners.

became advantageously known to Morellet as a man of science, and as one who had gratuitously and honourably conducted a useful work in France. His predominating taste thus continued, as in former times, its influence, and was still a connecting link between him and new and old friends. On this and many other occasions he proved the truth of what has been asserted, that no effort is ever lost; his exertions at Lyons in 1772, after an interval of thirty years, now becoming of unexpected advantage to him and to his family at Paris.

The Abbé Morellet, on renewing his acquaintance with my father, found, that the vivacity, frankness, and candour of both their dispositions made them particularly suited to each other. An intimacy immediately commenced. It began, however, by a warm debate at a literary breakfast, where some argument arose about the exact meaning of the words *perfectibilité* and *perfectionnement;* when, to the astonishment of the company, my father, a stranger, a foreigner, speaking the French language but imperfectly, maintained his opinion against *le doyen de la littérature Françoise.* Instead of being offended, the venerable and amiable Abbé wrote him a note early the next morning, acknowledging, that, upon examining the authorities, he found himself in the wrong. The high character which M. Morellet bore, not only for learning and sound judgment, but for the courage and uncommon consistency of his conduct through the whole of the Revolution, gave him extraordinary influence in that society in Paris, to which we were ambitious of being introduced, and which was composed of all that remained of the ancient men of letters, and of the most valuable of the nobility:—not of those who had accepted of places from Buonaparte, nor yet of those emigrants, who have been wittily, and too justly, described as returning to France after the Revolution, "*sans avoir rien appris, ou rien oublié.*"

Morellet answering for us, we were received at once into all that remained of that society, which the genius of Madame de Staël has so inimitably described. In that society we had the happiness of living during the whole time of our residence at Paris; and we felt the characteristic charms of Parisian conversation, the polish and ease, which in its best days distinguished it from that of any other capital. Most grateful my father felt at the time, and long continued to feel, for the kindness and confidence, with which we were treated by many, whom it would gratify, not merely my vanity, but my desire to do honor to his memory, here to enumerate. But knowing well their sentiments on this point, I refrain.

I may, however, without impropriety, be permitted to name those, who, by their stations and talents having publicly distinguished themselves, are in fact public characters. My father had not only an opportunity of becoming acquainted with the French literati, but with most of the celebrated men of science; particularly civil engineers, and men of invention and skill in mechanics.—M. de Prony, who was then at the head of *les Ponts et Chaussées,* shewed him in the best manner all that to a well-informed engineer was most worthy of notice in the Repository of that celebrated School; put him in the way of seeing every other invention, object, and person, in the mechanical and engineering department; but, above all, he felt grateful for M. de Prony's giving him so much of his own conversation, and for various indubitable proofs of private esteem and confidence.

Le Breton, then one of the principal officers at the Mint, is, as I am informed, at present at the Brazils: if so, there is little chance that the lines I am now writing should ever be seen by him; but, if they should meet his eye, may they convey to his heart the assurance, that my father, as long as he lived, remembered him with *gratitude*, and bequeathed that sentiment to his surviving family!

Berthollet, Montgolfier, and Breguet, gratified him by bestowing that gift, of which philosophers and men of science, occupied upon great objects, and independent of common society, best know the value; and which they, of course, cannot afford to bestow on any but those, who really interest them—their time. It was at this period, that we first became acquainted with our excellent friend, M. Dumont.

This gentleman, so well known by his conversational talents and his exquisite critical acumen, has entitled himself to the gratitude of the literary and political world in general, and of Englishmen in particular, by the successful pains he has bestowed in arranging, elucidating, and making known to the continent of Europe several valuable English works on legislation*—works, which, notwithstanding their depth of thought and extent of views, would never have acquired popularity,

* J. Bentham's, "*Traité sur la législation,*" and his "*Théorie des peines et des récompenses.*"

if they had not been rewritten in M. Dumont's clear and forcible style.

From the commencement of their friendship in 1802, my father continued to correspond with M. Dumont; and we owe much to his critical advice and sagacity in all our literary pursuits and publications.

One of the first things, that struck my father when he went to Paris, and one which he wished to see imitated in his own country, was the munificent liberality, with which all museums, public libraries, and national institutions are, or *were* open, both to foreigners, and to the natives of the country, without any paltry fees or petty patronage*

Much valuable information is given, in the most agreeable manner, by the learning and talents of those at the head of the several museums and institutions, to individuals who have particular recommendations, or to those who have any title to that regard due, and in

* " But what the flow'ring pride of gardens rare,
" However royal, or however fair,
" If gates, which to access should still give way,
" Ope but, like Peter's Paradise, for pay?
" If perquisited varlets frequent stand,
" And each new walk must a new tax demand?
" What foreign eye but with contempt surveys?
" What muse shall from oblivion snatch their praise?"

Paris always paid, to all who are of the republic of letters.

As a member of the English Royal Society, my father was invited to the honors of the sitting at the French Institute; but wishing to claim this privilege for the Royal Irish Academy, he begged to be permitted to take his seat in the Institute as a member of that Academy, and declined taking it as F. R. S. This was perhaps considered as a waste of patriotic politeness by some English gentlemen then present; but on his return home he was thanked for his conduct by Mr. Kirwan, (President of the Royal Irish Academy,) and by those members of that Academy, who were conscious, that their own fame had deservedly reached to foreign countries. Frequently at Paris he had the pleasure of answering inquiries about our celebrated countrymen, Dr. Young, and Dr. Brinkley (Professor of Astronomy at the University of Dublin), whose names and talents are held in the highest esteem by all men of science on the continent.

In Paris there is an institution resembling our London Society of Arts; "*La Société d'Encouragement pour l'Industrie Nationale;*" of this my father was made a member, and he presented to it the model of a lock of his invention. In getting this executed, he became acquainted with some of the working mechanics in Paris, and had an opportunity of observing how differently work of this kind is carried on there and in Birmingham. Instead of the assemblage of artificers in manufactories, such as we see in Birmingham, each artisan in Paris, working out his own purposes in his own domicile, must in his time "play many parts," and among these many to which he is incompetent, either from want of skill, or want of practice; so that in fact, even supposing French artisans to be of equal ability and industry with English competitors, they are left at least a century behind, by thus being precluded from all the miraculous advantages of the division of labour.

Of course every stranger who goes to Paris, whether he have, or have not, a taste for the arts, visits the great gallery of the Louvre. My father had always been to a certain degree fond of good pictures, but had never turned much of his attention to them, till he married a lady, who had a taste for painting. He now spent hours in the gallery of the Louvre; not to acquire the cant of connoisseurship, but with a desire to increase his knowledge of the arts, that he might enjoy more fully this subject of sympathy with one, who had bent all her pursuits to his, and who had turned her attention to those sciences, which he particularly liked.

Some people believe, that in marriage those who have different tastes and tempers suit the best. Difference between quick and slow tempers may perhaps be advantageous; because it may happen, that the faults of each may occur at different times, and that each person may bear with the other. But, if the temper be good, and well managed, the pleasure of sympathy will assimilate the tastes, without any danger of competition or jealousy. Indeed, it may be observed, that, where people are sincerely attached, they imperceptibly acquire each other's tastes.

My father had left England with a strong desire to see Buonaparte, and had procured a letter from the Lord Chamberlain (Lord Essex), and had applied to Lord Whitworth, our ambassador at Paris, who was to present him. But soon after our arrival at Paris, he learned that Buonaparte was preparing the way for becoming Emperor, contrary to the wishes and judgment of the most enlightened part of the French nation. Besides private conversations, from which he learned the sentiments of the best friends of France, his eyes were opened to the truth by an eloquent address to the First Consul, which had about this time appeared. The publisher of this address had been arrested; the author, Camille Jourdan, declared and surrendered himself; but so strong was public opinion in the author's favor, that Buonaparte, who ever warily watched the signs of the times, refrained from any attempt to punish the noble-spirited patriot*.

---

* A single passage from this pamphlet may shew the spirit of liberty, which previously to that time had existed in France.—

" Et pour prendre l'exemple qui nous touche de si près, nous avons un Buonaparte, nous en jouissons, nous l'admirons; mais qui nous l'a fait? Notre choix. Que seroit cet homme peut-être, si la gloire de ses aieux l'eût dispensé de la sienne; et puisque nous voulons le continuer dans les siècles, pourquoi ne pas nous confier à la cause puissante qui nous le donna? * * * * * Ainsi point de nom qui puisse enfler le cœur d'un chef, avilir celui de ses administrés, leur faire oublier à tous la relation primitive qui doit subsister entre eux. Serait-il vrai, par exemple, comme des écrits ont semblé l'indiquer, que quelques courtisans auraient médité de proposer l'adoption de quelque titre *d'Empereur*, *d'Empereur des Gaules*? Quoi, remonter ainsi dans la nuit des siècles, rechercher pour nous le nom que portaient nos barbares ancêtres quand ils étaient enchaînés par César, et trompés par les Druides! Nous faire abjurer ce doux et beau nom de la France consacré par tous les grands souvenirs de la monarchie, et par tous les triomphes de notre république! * * * * Voir ainsi un féodal *Empereur des Gaules* en tête de la charte

My father could no longer consider Buonaparte as a great man, abiding by his principles, and content with the true glory of being the first citizen of a free people; but as one meditating usurpation, and on the point of overturning, for the selfish love of dominion, the liberty of France. With this impression, my father declared, that he would not go to the court of an usurper. He never went to his levees, nor would he be presented to him.

I confess, that we used every argument and persuasion we could, to change his resolution; for we were exceedingly anxious, that he should seize this, perhaps the only opportunity that might occur, of seeing that extraordinary man. He persisted, however, in his resolution. We were disappointed and vexed at the moment; but I now look back with reverence to his motive, and to the steadiness of his resolution. When we used to regret, that he had seen Buonaparte only at the playhouse, opera, and review; he answered, that he could have seen no more of him, had he at-

---

libre des généreux Francs. Mais pourquoi s'arrêter à combattre de telles chimères? Et combien l'esprit national, d'accord avec le bon esprit de Buonaparte, doit nous rassurer contre de telles dénominations, si de lâches flatteurs osaient jamais les reproduire!"

tended the Consul's, or the Emperor's levee. It was impossible, he thought, from the trifling nature of what could be said at a levee, to judge of any thing more than the exterior. Even were he admitted to more private conversation, he did not imagine, that he could penetrate into the inward man. Each of the conversations of Buonaparte with various strangers and visitors, which have been repeated, exhibit only that *facet* of the mind, which it was the interest, or the humor of the moment, to turn outward; and it seemed an extraordinary degree of presumption or absurdity to conceive, that one practised in keeping his own counsels, and neither from nature, nor long habit, disposed to place confidence in his fellow-creatures, should throw his heart, if he have one, open to casual visitors.

My father had not the presumption to imagine, that in a cursory view, during a slight tour, and a residence of four or five months at Paris, he could become thoroughly acquainted with France. Besides, his living chiefly with the select society, which I have described, precluded the possibility of seeing much of what were called *les nouveaux riches*.

The few general observations he made on

French society at this time I shall mention. He observed, that, among the families of the old nobility, domestic happiness and virtue had much increased since the Revolution, in consequence of the marriages, which, after they lost their wealth and rank, had been formed, not according to the usual fashion of old French alliances, but from disinterested motives, from the perception of the real suitability of tempers and characters. The women of this class in general, withdrawn from politics and political intrigue, were more domestic and amiable; many wives, who had not formerly been considered as patterns of conjugal affection, having made great sacrifices and exertions for their husbands and families during the trials of adversity, became attached to them to a degree, of which they had not perhaps known themselves to be capable, during their youthful days of folly and dissipation.

With regard to literature he observed, that it had considerably degenerated. For the good taste, wit, and polished style, which had characterized French literature before the Revolution, there was no longer any demand, and but few competent judges remained. The talents of the nation had been forced by circumstances into different directions. At one time. the hurry and necessity of the passing moment had produced political pamphlets, and slight works of amusement, formed to catch the public revolutionary taste. At another period, the crossing parties, and the real want of freedom in the country, had repressed literary efforts. Science, which flourished independently of politics, and which was often useful and essential to the rulers, had meanwhile been encouraged, and had prospered. The discoveries and inventions of men of science shewed, that the same positive quantity of talent existed in France as in former times, though appearing in a new form.

The system of *espionage* in Paris at this time could not but strike every true Englishman with disgust and indignation. My father could scarcely be brought to believe in its existence, till he had well-attested facts produced to him, and till he perceived the suspicion, and excessive caution and constraint, which the system spread over general society. He never meddled in politics; and was, for his own sake, and for the sake of his friends, so prudent in conversation, and in the acquaintance he formed, that he had no apprehensions of attracting the censures of the police. Nevertheless, he was one morning surprised by an order to quit Paris in twenty-four hours, and the French territories in fifteen days. A letter of his own, written at the moment, gives a more lively description than I could of this slight affair.

TO MRS. CHARLOTTE SNEYD.

" MY DEAR C. SNEYD, *Paris, Jan.* 27, 1803.

" The only thin sheet of paper in the house had the words ' My Dear Sneyd' written at the top—by squeezing in a C. I make it serve for the beginning of a letter to you, my dear friend, to whom I have wished to write every day for this last week, without finding it possible; for in truth this has been an eventful week; and though the events of every week of my life, with my good will, should be detailed to you, with the utmost exactness; yet, during the moment that they pass, it is sometimes absolutely impossible, to write them down in order of battle; so that I often find myself passing over things that appeared worth relating, to come immediately to later circumstances, that fill my mind when I sit down to write to you.

" I had been lying in bed later than usual, to dissipate a cold, which had hung upon me for some time, and having just taken another *reef* of the blanket to keep my back warm, I heard a strange voice talking to Fanny in the next room; she came in and told me, that an officer of the police must speak to me instantly. I desired he might walk in; he appeared in his blue uniform, embroidered with green, and presented a warrant, which, as he informed me (in a manner sufficiently peremptory) required my immediate attendance. My being ill was not a sufficient excuse; I got up, and dressed myself *slowly*, to gain time for thinking—drank one dish of chocolate, ordered my carriage, and went with my *exempt* to the Palais de Justice. There I was shewn into a parlour, or rather a guard-room, where a man like an under-officer was sitting at a desk. In a few minutes I was desired to walk up stairs into a long narrow room, in different parts of which ten or twelve clerks were sitting at different tables. To one of these I was directed—he asked my name, wrote it on a printed card, and, demanding half a crown, presented the card to me, telling me it was a passport. I told him I did not want a passport; but he pressed it upon me, assuring me, that I had urgent necessity for it, as I must quit Paris immediately. Then he pointed out to me another table, where another clerk was pleased to place me in the most advantageous point of view for taking my portrait, and he took my written portrait with great solemnity, and this he copied into my passport. I begged to know who was the principal person in the room, and to him I applied, to learn the cause of the whole proceeding. He coolly answered, that if I wanted to know, I must apply to the *Grand Juge*. To the Grand Juge I drove, and after having waited till the number 93 was called, the number of the ticket which had been given to me at the door, I was admitted, and the Grand Juge most formally assured me, that he knew nothing of the affair, *but* that all I had to do was to obey. I returned home, and on examining my passport, found that I was ordered to quit Paris in twenty-four hours. I went directly to our ambassador, Lord Whitworth, who lived at the extremity of the town: he was ill—with difficulty I got at his secretary, Mr. Talbot, to whom I pointed out, that I applied to my ambassador from a sense of duty and politeness, before I would make any application to private friends, though I believed, that I had many in Paris, who were willing and able to assist me. The secretary went to the ambassador, and in half an hour wrote an official note to Talleyrand, to ask the why and the wherefore. He advised me in the mean time to quit Paris, and to go to some village near it—Passy or Versailles. Passy seemed preferable, because it is the nearest to Paris—only a mile and a half distant. Before I quitted Paris, I made another attempt to obtain some explanation from the Grand Juge. I could not see him, or even his secretary, for a considerable time; and when at length the secretary appeared, it was only to tell me, that I could not see the Grand Juge. 'Cannot I write,' said I, 'to your Grand Juge?' He answered hesitatingly—'Yes.' A huissier took in my note, and another excellent one from the friend who was with me—F. D. The huissier returned presently, holding my papers out to me at arm's length—'the Grand Juge knows nothing of the matter.'

I returned home, dined—ordered a carriage to be ready to take me to Passy—wrote a letter to Buonaparte, stating my entire ignorance of the cause of my *déportation,* and asserting, that I was unconnected with any political party. F. D. engaged, that the letter should be delivered; and Mrs. E. and Charlotte remaining to settle our affairs at Paris, I set off for Passy with Maria, where my friend F. D. had taken the best lodging he could find for me in the village. Madame G. had offered me her country house at Passy; but though she pressed that offer most kindly, we would not accept of it, lest we should compromise our friends. Another friend, Mons. de P , offered his country house, but for the same reason this offer was declined. We arrived at Passy about ten o'clock at night, and though a déporté, I slept tolerably well. Before I was up, my friend Mons. de P. was with me—breakfasted with us in our little oven of a parlour—conversed two hours most agreeably. Our other friend, F. D., came also before we had breakfasted—and just as I had mounted on a table, to paste some paper over certain deficiencies in the window, enter M. P—— and Le B * * * * n.

"'Mon ami, ce n'est pas la peine!' cried they, both at once, their faces *rayonnant de joie.* 'You need not give yourself so much trouble—you will not stay here long. We have seen the Grand Juge, and your detention arises from a mistake. It was supposed, that you are brother to the Abbé Edgeworth—we are to deliver a petition from you, stating what your relationship to the Abbé really is. This shall be backed by an address signed by all your friends at Paris, and you will be then at liberty to return.'

"I objected to writing any petition, and at all events I determined to consult my ambassador, who had conducted himself well towards me. I wrote to Lord Whitworth, stating the facts, and declaring, that nothing could ever make me deny the honor of being related to the Abbé E. Lord Whitworth advised me, however, to state the fact, that I was not the Abbe's brother."

P * * * * * and Le B * * * * * * drew up a memorial in my father's favor, and presented it to the Grand Juge at a public audience. His reply was a question—"Is not this Englishman brother to the Abbé E.?"—Pretending to be satisfied when he was apprised of the contrary, he said, that the affair was all a mistake, and that the gentleman might return to Paris. No direct answer was received from the first Consul; but perhaps the revocation of the order of the Grand Juge came from him. We were assured, that my father's letter had been read by him, and that he declared he knew nothing of the affair; and, so far from objecting to any man for being related to the Abbé Edgeworth, he declared, that he considered him as a most respectable, faithful subject, and that he wished he had many such. In the same spirit, and with the same expressions, he had about this time given a place to the only surviving counsellor of Lewis the XVIth, who had defended the King on his trial.

In the evening of the second day of my father's *banishment* from Paris, our friends informing Mrs. Edgeworth of the permission granted him to return, she came to Passy for us at seven o'clock in the evening. Late as it was, when he got to Paris, he stopped at the English ambassador's hotel, to tell him the result of the business.

At a public court dinner, at which Regnier, the Grand Juge, was present, some days after this affair, one of our friends spoke of it, and questioned him as to his *real* reasons. He declared he had none; but excused himself by saying, that Paris was too full of strangers, and that he had general orders, to clear it of *la lie du peuple étranger;* to which our friend replied, that "M. le Grand Juge should however distinguish between *la lie* and *l'élite du peuple.*"

By every playful and every serious expression of kindness, our friends endeavoured to efface from our minds any disagreeable recollection of what had happened. As to personal inconvenience or transient anxiety, occasioned by an eight and forty hours banishment, it was infinitely overbalanced to my father, by the means it afforded him of seeing the real regard felt for him by those, whom he most esteemed in France. It gave us also an opportunity of seeing the character of the French in the most amiable point of view, and confirmed his former belief in the steadiness and substantial nature of their friendships.

The memorial, which our Parisian friends drew up, to present to the Grand Juge, stated, " that my father was a man of letters, that we were authors of a work on education, well known in France—that he had lived, ever since he came to Paris, with literary society, totally unconnected with politics." Some kind and highly gratifying expressions were added: several celebrated names of the highest respectability were subscribed. After the business was over, the memorial was put into my father's hand, and has been, and will be carefully preserved by his family, as a testimony of the steadiness of our Parisian friends. It should be observed, that all this happened just after the explosion of what was called *the infernal machine,* when Buonaparte had been put in fear of his life, and when he was peculiarly suspicious of foreigners, and jealous in an extraordinary degree of the English, and of all who liked or frequented their society. Our friends were well aware of the *disagreeable* consequences to themselves, that might have ensued from exciting his displeasure. At the moment when they offered their country-houses, and their active assistance and support, they did not know, that the whole affair originated with the Grand Juge; they had, on the contrary, reason to believe, that it came from Buonaparte himself. Therefore there was the more hazard and merit in their kindness.

As the spring was advancing, my father was pressed by his friends to prolong his stay; they assured us, that a summer at Paris is more delightful, than can be imagined by any, who have not experienced its charms. The society there pleased him so much, he saw so many means of instruction and amusement for his children, and of improvement for the grown and growing up part of his family, that before this affair occurred, he had more than half resolved to send to Ireland for them, and to take a house at Paris for two years. Previously to his being ordered to leave Paris, he was in treaty for a house formerly belonging to Garat, in a charming situation, near les Jardins du Luxembourg. But, notwithstanding the proofs of disinterested and steady attachment, which he had received from his friends, and notwithstanding all the allurements of Parisian society; an Englishman, with British feeling and the spirit of liberty within him, could not, after all that had passed, desire to reside under a government, where he might, innocent of all offence, be seized by a petty officer of police, and without examination, trial, or reason given, be ordered to quit the capital in four and twenty hours. He was no longer tempted by Garat's house in the Luxembourg gardens. He was prudent and decided—had he been otherwise, we might all have been among the number of our countrymen, who were, contrary to the law of nations, and to justice and reason, made prisoners in France at the breaking out of the war. We were fortunate in getting safe to free and happy England a short time before war was declared, and before the detention of the English took place.

My eldest brother had the misfortune, to be among those who were detained. His exile was rendered as tolerable as circumstances would permit, by the indefatigable kindness of our friends the D'*******s. But it was an exile of eleven years—from 1803 to 1814—six years of that time spent at Verdun! Had this happened to any younger brother, who had his fortune to make by his profession, it must necessarily have been ruin to his prospects for life. For an eldest son, there was another sort of danger, the danger of his being drawn into bad company and dissipation, by mere want of agreeable employment. Under such circumstances, and in such a place as Verdun, and during such an exile from his native country, his home, and friends, there was great hazard, that his domestic affections and his habits of life might be changed. This was to be a trial of the efficacy of his education.

CHAPTER XIV.

When we were returning from Paris, we heard such an account of the declining health of my brother Henry, who was then at Edinburgh, as determined my father to go immediately to Scotland to see him, and to bring him home with us to the milder climate of Ireland.

We went to Scotland in the spring of 1803; found Henry's health and spirits better than we had expected; his kind and skilful physician and friend, Dr. Gregory, gave us hopes, that, with care and caution, he might escape threatening consumption. My father was ever ready to hope, and his confidence in the judgment and care of Dr. Gregory were so far justified; for Henry recovered from this attack, and we flattered ourselves, that by care he might be long preserved. He mended rapidly while we were at Edinburgh; and this improvement in his health, added to the pleasure his father felt in seeing the interest his son had excited among the friends he had made for himself at Edinburgh—men of the first abilities and highest characters, both in literature and science—whom we knew by their works, as did all the world; with some of whom my father had had the honor of corresponding, but to whom he was personally unknown. Imagine the pleasure he felt at being introduced to them by his son, and in hearing Gregory, Alison, Playfair, Dugald Stewart, speak of Henry, as if he actually belonged to themselves, and with the most affectionate regard.

From the time he came to Edinburgh, to the hour he left it, Henry was received at Lothian House, where Mr. and Mrs. D. Stewart then resided, as if he had been one of their own family. There all his holidays, all the hours that could be spared from study, were delightfully spent; and there, in health and in sickness, he received all the counsel and sympathy, and all the tender care, which the best of parents could have bestowed. As one of those friends said to me—"this pure and innocent being was, while in Edinburgh, watched over with constant and unwearied affection."

May I? Yes—I *must* be permitted to name Mrs. Dugald Stewart, for whom he always felt the gratitude of the most affectionate son to the kindest of mothers. To her maternal care, and to Mrs. Alison's, we owe it, that Henry got through two severe seasons in Scotland, and that a few years longer of his life were preserved.

We spent some weeks with him, and among his friends at Edinburgh, in delightful society. The evening parties at Lothian House appeared to us (though then fresh from Paris) the most happy mixture of men of letters, of men of science, and of people of the world, that we had ever seen. And here, while we enjoyed the powerful display of talents, and the pleasures and profits of conversation, we felt a sense of security and satisfaction, from knowing, that beyond and above these there existed British liberty, and domestic virtue and happiness.

We left Edinburgh with sentiments of public admiration and private gratitude, regretting, that we must quit such society, and part from such friends. Dr. Gregory advised, that no delay should be made in removing Henry, who was now equal to the journey, to a milder climate.

On our journey homewards, in passing through Scotland, we met with much hospitality and kindness, and much that was interesting in the country and in its inhabitants. But the circumstance, that remains the most fixed in my recollection, and that which afterwards influenced my father's life the most, happened to be the books we read during our last day's journey. These were the lives of Robertson the historian, and of Reid, which had been just given to us by Mr. Stewart. In the life of Reid there are some passages, which struck my father particularly. I recollect at the moment when I was reading to him, his stretching eagerly across from his side of the carriage to mine, and marking the book with his pencil with strong and reiterated marks of approbation. The passages relate to the means, which Dr. Reid employed to prevent the decay of his faculties as he advanced in years; to remedy the errors and deficiencies of one failing sense by the increased activity of another; and try the resources of reasoning and ingenuity to resist, as far as possible, or to render supportable, the infirmities of age.

The philosophic and benevolent biographer expresses a hope, that, recording the success of these endeavours of Dr. Reid's will encourage others, and lead both to the improvement of the human understanding, and to the prolongation of the happiness derived from the exercise of our faculties. In one instance this wish, as I can attest, has been accomplished. My father never forgot this passage, and acted upon it years afterwards.

In the latter end of the year 1803, and in 1804, after our return to Edgeworth-Town, he was again intent upon that object, in which he had formerly met with disappointments, sufficient to have discouraged the ardor, and disgusted the patriotism of most other men. But convinced as he was, that it would be peculiarly advantageous to Ireland, he pursued it with unconquerable perseverance.

When Lord Hardwicke was Lord Lieutenant of Ireland, government at length desired the establishment of a telegraph in this country, and my father was employed to form a line of communication from Dublin to Galway. Captain Beaufort of the Navy, (his friend and brother-in-law) not being at that time in active service, engaged to assist in this undertaking, both from private friendship, and from a belief, that it would be beneficial to the country. He would not accept of any pecuniary remuneration, and devoted to this object two years of his life in unremitting, zealous exertion.

A line of telegraphs from Dublin to Galway was completed, temporary guard-houses were built at the requisite stations, and a telegraphic corps was formed from those of my father's yeomanry corps, and others of his tenantry, whom he had judged fit for the purpose. They had some additional pay, and their expenses of living at different stations were defrayed by government. They conducted themselves invariably well, during the whole time they were employed, and shewed all the steadiness and intelligence, that could be desired. Telegraphic messages and answers from Dublin to Galway were transmitted in the course of eight minutes, in a public experiment for the Lord Lieutenant; and his Excellency's approbation was graciously expressed. Every one seemed perfectly convinced of the utility, and satisfied of the efficiency and success of the establishment. The telegraphs being portable, they could be erected or taken down in a few minutes, and the whole line might thus be removed into any direction, that the will of the commander in chief, or the exigency of the moment, should require—two men could, with ease, carry the whole paraphernalia of each station upon their shoulders.

To shew how rapidly the stations might be rendered secure from any sudden attack, and available for other purposes in disturbed parts of the country, my father proposed to fortify one of them, suggesting, that the experiment should be made on the hill of Cappa, the third station from Dublin, and a place where the advantage of having a military post had been sufficiently manifest during the rebellion of 1798. Government empowered him to carry this plan into execution. He obtained possession of a ruined stone windmill on that hill, and at a wonderfully small expense rendered it proof against any species of hostility, except that of cannon, or blockade.

The views of Buonaparte, however, seemed to be directed to other shores. The alarm of invasion gradually subsided, and the telegraphs were consigned to the care of the ordinary military established in the country. My father and his friend being diplomatically thanked for their exertions, the latter returned to the more active pursuits of his profession, and we again rejoiced in the undisturbed peace of the country.

I cannot quit this subject, without expressing what I know were my father's feelings towards the Irish military secretary, Sir Edward Baker Littlehales*, who in all this business shewed the most liberal and enlightened views for Ireland. On every occasion he treated her country gentlemen, and all who

* Now Sir E. B. L. Baker.

had business to transact at the Castle, with that politeness, temper, and uniform propriety, which in his station is equally becoming and conciliating. His peculiar attention and kindness to my father I attribute in a great measure to his knowledge, that Mr. Edgeworth was really an honest, independent gentleman, anxious to contribute as far as he could to the advantage of this country. In this public point of view I may therefore publicly offer him these thanks—the more willingly, because he has now retired from office.

During the year when my father was engaged in establishing this telegraph, he always found time to attend to the education of his children, and in some way or other turned to account for their improvement whatever active employment engaged his attention. About the time when he had completed this telegraphic business, one of my younger brothers, whom he had educated, was nearly ready to enter College.

When I was writing this page, (July, 1818,) this brother was with me; and when I stopped to make some inquiry from him as to his recollection of that period of his life, he reminded me of many circumstances of my father's kindness to him, and brought to me letters written on his first entrance into the world, highly characteristic of the warmth of my father's affections, and of the strength of his mind. I regret more than I can express, that these cannot be given to the public. But every one, who is called upon to write the lives of those recently departed, must know and bear these struggles between love and admiration for those who are no more, and conscientious regard to the feelings and interests of the living.

One general observation, which may, perhaps, be useful to the public, I may make, while the conviction is full and strong on my own mind, that a father's confiding kindness and plain sincerity to a young man, when he first sets out in the world, make an impression the most salutary and indelible. When his sons first quitted the paternal roof, they were all completely at liberty; he never took any indirect means to watch over, or to influence them; he treated them on all occasions with entire openness and confidence. In their tastes and pursuits, joys and sorrows, they were sure of their father's sympathy; in all difficulties or disappointments, they applied to him as their best friend, for counsel, consolation, or support; and the delight, that he took in any exertion of their talents, or in any instance of their honorable conduct, they felt as a constant, generous excitement. In short, he conducted himself to his family uniformly upon the principles, which he has recommended in his writings; and this happily for a long course of years, and with a succession of the children of different mothers. But repeatedly he was doomed to see the fairest blossoms of talent blasted by disease, and the most highly cultivated, and the most valuable fruits of education perish, almost at the moment when they attained to perfection beyond his fondest hopes.

It was not the health of my brother Henry, for which we had been most alarmed, that first gave way; but that of one of his sisters, for whom we had never had any apprehensions; one who had appeared the image of health, Charlotte, that young person, formerly described as being "fresh as a rose." Soon after her return from the Continent, her health changed and declined; but, as she did not resemble either of her sisters, Honora or Elizabeth, who died of consumption, this difference long gave flattering security.

In the autumn of 1806, however, the symptoms of pulmonary consumption appeared too plainly, and were too quickly and strongly confirmed, to leave any further hope.

She died the ensuing spring, (April, 1807,) in her twenty-fourth year. The last act of her life, and the last connected sentence she uttered, was dictating to Mrs. Edgeworth, in words of tender esteem and gratitude, a bequest of a separate property she possessed, to the children of the beloved sister, (Emmeline,) to whose care of her early youth she owed so much.

Charlotte had been admired wherever she had been seen, abroad or in her own country; wherever she was known, she had been loved, and most loved by those who knew her best. Her character and understanding had been developed, and she had been placed in circumstances, where she had opportunities of fully proving both her steady judgment, and her amiable disposition. So that beyond the mere promise of early youth, there had been certainty and proof of excellence, which had endeared her peculiarly to her father, and he had looked forward to her for established years of happiness. She was not regularly beautiful; but, she had an uncommonly prepossessing and ingenuous countenance, with a peculiarly serene and happy appearance. She had engaging modesty of manners, without any of that embarrassment and awkwardness, which sometimes accompany timidity. She early shewed a talent for drawing, which was, of later years, cultivated by the instructions of Mrs. Edgeworth ; but never with any view to exhibition, or any hope, but that of giving pleasure to her friends, and of securing agreeable occupation for herself. I have, however, ventured to have a few of her sketches engraved, that the public may, from these specimens, form in one particular some judgment of her talents; and that it may not be supposed, that what has been said of her in other respects is the mere language of partiality*.

It is but just to the father, whose life I am writing, to give, as far as it is in my power, every proof of the promise of excellence in those of his children, whom he had the misfortune to lose.

The loss of Charlotte was to be followed by that of her brother Henry, though not immediately; yet, assuredly, the one calamity led to the other. His health, which had never been strong, was much affected by her illness, and received a shock at her death, which it never afterwards recovered. He made the most laudable efforts to pursue his studies, and to prepare himself for his profession of medicine; having the most ami-

* See Appendix.

able ambition to do justice, as he said, to his education, and honor to his family. He went to London, took his degree, went through the London *Hospitals, exerted himself beyond* his strength—all in vain. Two years after his sister's death, his health failed. He was obliged to go abroad, to Madeira: intervals of amendment for some time flattered us, that he would recover, but these hopes were false; he returned to England, and died at Clifton.

CHAPTER XV.

An opportunity of being usefully employed for the public again roused my father's mind to exertion. In 1806, under the administration of the Duke of Bedford, who, happily for Ireland, was then its Viceroy, a board of commissioners to inquire into the education of the people of Ireland was appointed by His Grace. My father had claims to the honor of being of this board, as having brought into the last Irish parliament a bill for the better education of the people, as having resided long in Ireland, and as being a person, whose principles of toleration had been manifest in his conduct, and whose zeal for the improvement of education was known by his writings. These claims were felt; and in the most handsome manner, without any direct or indirect application on his part, he was appointed one of the commissioners. They received no salary. Their board lasted from 1806 to 1811, when their last Report was finished, and sent to the British Parliament. These reports have been printed by the House of Commons; but as they may not reach the hands of many, who have not the means of obtaining parliamentary papers, it may be useful to lay fully before the public that letter to the Archbishop of Armagh, Primate of Ireland, on which, as it has been observed by the Duke of Bedford, the best part of the most liberal report of the Board of Education was founded.

The reader will find it in the Appendix.

Another letter of my father's, printed in these reports, is also given in the Appendix, as containing a short, satisfactory view of the Charter Schools of Ireland, and some hints for their improvement. These were addressed to the Primate, in the intervals between the periods when the board met, and they were usually read aloud by the secretary at the succeeding meeting. Thus time was saved by the arrangement of sentiments in writing, and the sum of the opinions of individual members obtained at once, and subjected to discussion at the meetings of the commissioners. This Board was attended regularly, at every summons, by all the members, who assembled for that purpose in Dublin, though many lived at distant parts of the country. To my father this was frequently inconvenient, but he never missed any one meeting. The Primate was indefatigable in his zeal and patience, and substantial good resulted from the inquiries of the commissioners. Some abuses were detected and reformed, and the public obtained a clear view of the present state of the Charter Schools, and other schools in Ireland, with suggestions for the best means for their improvement. But, above all, the great point gained was the assurance from this commission, composed of many of the dignitaries of the Church, with the highest at their head, that the system of proselytism is abandoned, and that it is their wish to proceed in the most liberal manner towards the Catholics of Ireland, in the further improvement of the education of the people.

A private good, to my father, on which he set the highest value, resulted from the manner in which he conducted himself at this Board. He obtained the approbation, and, during his life, ever afterwards enjoyed the esteem and peculiar regard of the present Primate of Ireland*.

About this time Lord Selkirk applied to my father for information respecting the situation and dispositions of the lower class of people in Ireland. The following extracts from his letter in reply give a full view of his opinion of them, and mark the principal alterations he observed to have taken place within the last ten years.

TO THE EARL OF SELKIRK.

* * * * * * "The progress of agricultural improvement in the greatest part of Ireland has been very different from what has happened in Scotland. There, the arable lands are converted into pasture, because feeding sheep and cattle is more advantageous to the landlord, than feeding the miserable and unprofitable children of cottagers. Here, the rich pastures are dug up to produce potatoes for families, whose wretchedness and sloth do not impoverish the owner of the soil; because he receives more for his land, when employed in this sort of tillage, than when it fattens beef and mutton: as long as the lazy inhabitant of a cabin can provide for his family, "meat, fire, clothes," he will not be tempted from that dear hut, his home.

"It is in vain that we despise his sordid content; turf supplies fire and smoke, so that he has warmth enough to compensate for the insufficiency of his cabin; potatoes not only supply himself and his family with food, but afford a redundancy—sometimes for a pig, and always for a beggar; and as to clothes, the difference between rags and a whole coat is not much regarded amongst his neighbours.

---

* The Hon. Dr. Wm. Stewart, son of the Earl of Bute.

" It is true, that this submission to avoidable poverty is becoming less common and less creditable; but whenever more energy is exerted, it is near home: wicker chimneys, a pane of glass, and a cabbage-garden, are becoming every day appendages to the cabin; and the astonishing increase of white stockings and cleanliness among the women necessarily compels the men to labor for the purchase of these new luxuries: but still these luxuries, and these very exertions, fasten the cottager to the soil; and I say, from my own knowledge, that the dreams of emigration disturb the lower classes in my neighbourhood much less than formerly. That such an order of things cannot last indefinitely, is not to be disputed; the sons of cottagers expect, as they grow up, to divide with the aged father the small farm which he possesses, because they are better able to cultivate it than he is, and because the son considers something now due to him for his labor during the profitable years of his early youth: a tacit acquiescence in these claims is nearly universal; and where the sons do not go into the army or to sea, or to some trade or manufacture, the land is subdivided, till it ceases to provide for another marriage, and then necessity drives the young brood from home; the landlord loses his rent for a year or two; he lets it to new adventurers at an increased value, which repays his former loss.

" New industry arises with new hopes; useful luxury obliges men to look forward. Fathers use their best endeavours to give their children the elements of education; the children perceive, that this education is not an evil inflicted by old age on childhood; but they are early sensible, that knowing how to read and write prepares them for situations something above that of day-labourers or wretched cottagers.

" In a school for the sons of peasants, and in one for the sons of gentlemen, the looks, and obvious feelings of the scholars are very different: the poor generally are intent upon their business, because it is their business—the rich often consider their tasks as needless impositions, that do not necessarily lead to any future obvious advantage, and of course employ as little exertion as possible in their accomplishment.

" I refer to facts—Latin and Greek are out of the comparison—but I rely upon the event of any trials, that may be made upon boys of the higher and lower classes in Ireland, in which I am certain it will be found, that not only the common, but the higher parts of arithmetic, are better understood, and more expertly practised by boys without shoes and stockings, than by young gentlemen, riding home on horseback, or in coaches, to enjoy their Christmas idleness. This application is abundantly rewarded: boys well taught become clerks if they remain at home, and sergeants if they go into the army: and for the honor of human nature, and in support of those who maintain the moral advantages of education, the gratitude of these young Irishmen is so common, as scarcely to excite praise.

" I have seen numbers of their letters, full of true filial affection; and frequently containing exhortations to their friends, never to neglect the education of the brothers and sisters, whom they had left behind them.

" The views that urge the people to this mode of conduct, with respect to their children, are in no way connected with plans of emigration—they look forward in the real hopes of being supported in their old age by children, whom they breed up when young; and in common, I believe, with the poor of Scotland, they look abroad for provision for their sons, but not to distant colonies, for the preservation of family independence and the honor of their name.

" The garden culture of Ireland renders each family in some degree independent, as to mere subsistence: but at the same time it prevents the growth of corn, and retards agricultural improvement. It creates a nursery for soldiers, but not for emigrants.

" Nothing can be more clearly demonstrated, than what you assert with respect to clanship, and the decay of that military spirit, which formerly enabled gentry to lead bands of real soldiers to the army; this spirit, as you have shown, does not prepare them for manufacturers, or for common recruits: therefore till time has obliterated their ancient feelings and habits, there is no resource for increasing population, but new settlements.

" Our redundancy, when it is felt, must submit to war and pestilence; famine will probably, in this country, yield precedence to pestilence, as the sustenance of human life, upon the wretched scale on which it is measured in Ireland, cannot fail, till its population is nearly doubled.

" Behold! two sheets in answer to the first four lines of your Lordship's paper. * * * * * * * * * *

" A great portion of this soil belongs to absentees, who have long since felt, that agents would not, or could not, collect the rents of cottagers; and every man, whom business, or pleasure, has drawn away from actual residence on his own estate, has found the same. The number of middlemen have lately given place to actual occupiers of small farms, from the circumstance which you allude to. Some savings were made by these people, which enabled them to take and stock farms of twenty or forty acres; and the larger farmers have been obliged, either to employ their capitals in improving smaller portions of ground, or in some other business.

" You observe, that within a very few years a great change has taken place, as to the occupancy of land; particularly since the year 80, when that judicious inquirer, Mr. A. Young, visited this country: but within these last five years a still greater change has occurred; the prodigious and sudden increase of prices for every article that land produces, and the temporary dearth of whiskey, accumulated considerable sums in the hands of middling tenants; with these savings, they bid for more land, and at a higher rent than middlemen would risk, and to the present time they have succeeded; but at the moment (1806) whilst I write, a deplorable and alarming change has shaken this system to its foundation. Yarn and cattle have fallen so much in price, as to leave no means in the power of the tenants to pay their high rents, but from their former savings; and the probable consequences will be a vibration as violent on the other side—that panic will deter farmers from taking land, and folly will yield, on the one hand, to imposition, or, on the other, will obstinately refuse mercy to all debtors.

" But still the army and navy will carry off our spare hands, and till emigration has become a systematic outlet for the young adventurer, I do not think, that it will become common from Ireland. Many failures amongst those, who emigrated from this country to America, have long since discouraged this experiment; which, after the close of the American war, was considered as a sure means of acquiring wealth and happiness.

" Upon the whole, I do not think that the present Irish are disposed to emigrate, nor are they fit for emigration; they are sanguine, but not persevering. From the highest to the lowest, their imaginations are easily dazzled, but they seldom know how to keep their eyes steadily fixed on one object. They have not that sense of family pride, that linked the Scotch together in tribes, which one common interest pervaded—though social, they are not united; though ready to combine in any sudden enterprise, they never submit to that subordination, which can alone ensure success: at the time of the rebellion, it was soon found, that every man was, or wanted to be, a captain.

" Were a man in whom they had entire confidence, who was far their superior in rank, fortune, and estimation for courage and abilities, to lead two or three thousand of the people of Ireland to a colony, where the land was fertile, and the climate temperate, I make no doubt of his success; for notwithstanding all I have said, there is no people, with whom I have ever been conversant, whom I would sooner chuse for such an experiment, under *such circumstances*. But the expense of the voyage, and of necessaries, could not be found by the people themselves; and to a large amount, nothing but national funds could suffice. These could not be obtained, and without them, I know but one man in the world that could successfully establish a colony of Irish in America.

* *

"We have at present a commission sitting for the purpose of forming a plan for educating the people of this country: whoever contributes to enlighten us on this subject will be a real benefactor to Ireland; and whether the posterity of its present inhabitants seek a distant climate, or remain to cultivate their native soil, if they can be taught sobriety, habits of order and obedience, and the proper use of their understandings, in guiding their conduct to their own happiness, a real, permanent, and increasing benefit, will be conferred on the country.

"I send your Lordship an abstract of part of the statistical survey, which I have made of this county. It will be published as soon as I can get a map finished to which it refers. It is a view of a district, where I had an opportunity of obtaining the most accurate account of what I wished to know; of the rest of the county I have procured returns, which are, I believe, as accurate as usual; but what I send to you I rely on."

During the years 1807-8 and 9, my father was engaged in some experiments on wheel carriages, and in literary occupations, especially in writing the narrative of his life. In the autumn of 1809 he was suddenly interrupted in that narrative by a severe illness. It seemed to be a mixture of bilious and nervous fever. By whatever name it was to be called, it was alarming, and it weakened his constitution more than any former attack. He had not recovered, when he heard of a new object of public utility, on which his ardent mind, in spite of all bodily sufferings, became happily intent.

Commissioners were appointed, to examine into the nature and extent of the bogs of Ireland, and to determine whether they could or could not be reclaimed. One of these commissioners, a private friend, was particularly anxious to engage my father's active assistance in the business; but after some letters had passed, from which he had no idea, that my father was seriously ill, happening to come for a day to Edgeworth-Town, and seeing the state to which he was then reduced, his friend imagined it would be not only dangerous, but impossible for him, especially at his advanced time of life, to undertake an employment, where he must be exposed to great bodily fatigue. My father, however, said he could only die, and that he would rather die doing something, than doing nothing. Besides its being a great national object, he was also interested in this business on account of one of his sons, who was intended for a civil engineer; and this undertaking would, he thought, afford opportunity of advantageously conducting his son's professional education. After privately remonstrating in vain against some articles in the instructions of the board to their engineers, which he thought would lead to great and needless expense, he accepted the employment, being convinced, that the undertaking, even though it might not be managed in the most economical or best manner possible, would, nevertheless, pay the public by its ultimate utility. He undertook the charge of a district, containing above thirty-four thousand five hundred English acres, which, according to his instructions, were "to be "surveyed, levelled, and bored to the hard "ground, at lines parallel to each other, at "the distance of a quarter of a mile." However he might previously remonstrate as an adviser, from the moment he acted as one under command, no man could obey more scrupulously, or labor more zealously.

In pursuance of the instructions of the Board, which were, to establish means of verifying the maps delivered by their engineers; Mr. William Edgeworth, with scientific accuracy, laid down a base for a series of triangles for the trigonometrical survey of this district; and with the assistance of his friend Dr. Brinkley, (Professor of Astronomy in Ireland) and with an excellent transit instrument of four feet length, he determined a meridian line; he also ascertained the latitude with a repeating circle of sixteen inches diameter, constructed by Troughton.

It was late in the year, and the weather unfavorable. In laying out and verifying the work of the surveyors employed, my father was usually out from day-break to sunset—often fifteen hours without food, traversing on foot, with great bodily exertion, wastes and deserts of bog, so wet and dangerous, as to be scarcely passable at that season, even by the common Irish best used to them.—In these bogs there frequently occur deep holes, filled with water of the same color as the bog; or sometimes covered over with a slight surface of peat heath or grass, called by the common people *a shaking scraw.*

In traversing these bogs, a man must pick his way carefully, sometimes wading, sometimes leaping from one landing place to another, chusing these cautiously, lest they should not sustain his weight; avoiding certain treacherous green spots, on which the unwary might be tempted to set foot, and would sink, never to rise again. It may be conceived what fatigue and anxiety must attend these operations to a person unused to them; especially to one in my father's habits of life, and brought up, as the reader may remember that he describes himself to have been, too tenderly.

He was now in his 65th year. To our astonishment his health not only endured these exertions, and the hardships to which he exposed himself, but improved under them, and was reestablished. This confirmed him in the belief of his favorite doctrine, that the mind can sustain the body. After nearly a twelvemonth's hard work, his part of the undertaking was completed, and his report was given to the Board. It is published among the parliamentary reports for 1810. His general opinion was, that the improvement of the vast tract of bog, which he examined, was not only practicable, but that it would prove highly and quickly profitable to individuals, and advantageous in every point of view to this country, and to Great Britain.

The public reports of the other engineers employed, and the private conversation he had with them, confirmed him in this belief. But he was aware, that his and their opinion, evidence, or reasoning, would not, and ought not to be sufficient alone, to impress this conviction upon the public. Therefore his report strongly urged, "that some public experi-"ment should be tried, that might demon-"strate to the nation, either that the scheme

" of improving the bogs of Ireland is practi- " cable or hopeless."

The chief obstacles are the want of capital, and the danger of litigation. The want of capital in the immediate proprietors, or in this country in general, appeared to him of no irremediable consequence; because if once the public mind were satisfied of the safety and advantage of the speculation, English capital would probably flow to Ireland for this purpose, and large companies would be formed, to carry on any undertaking, which promised high and secure profit. The danger of litigation he considered as a greater obstacle. Of this he had experience in his own case. He wished to undertake the improvement of a large tract of bog in his neighbourhood, and for this purpose desired to purchase it from the proprietor: but the proprietor had not the power or the inclination to sell it. My father, anxious to try a decisive experiment on a large scale, proposed to rent it from him, and offered a rent, till then unheard of for bog land. The proprietor professed himself satisfied to accept the proposal, provided my father would undertake to indemnify him for any expense, to which he might be put by future lawsuits concerning the property or boundaries of this bog. He was aware, that, if he were to give a lease for a long term, even for sixty years, this would raise the idea, that the bog would become profitable; and still further, if ever it should be really improved and profitable, it would become an object of contention and litigation to many, who might fancy they had claims, which, as long as the bog was nearly without value, they found it not worth while to urge. It was impossible to enter into the *insurance* proposed, and consequently he could not obtain this tract of bog, or further prosecute his plan. The same sort of difficulty must frequently recur. Parts of different estates pass through extensive tracts of bog, of which the boundaries are uncertain. The right to cut the turf is usually vested in the occupiers of adjoining farms; but they are at constant war with each other about boundaries, and these disputes involving the original grants of the lands, hundreds of years ago, with all subsequent deeds and settlements, appear absolutely interminable.

It is said, that they might be settled by the intervention of *a commission of boundaries.* But supposing the original title of the proprietors to be established, further difficulty, depending on wills and settlements made by the tenants, and leases to under tenants, create inextricable entanglements, which must prevent the purchase or sale of these bog lands. No hand but that of the legislature can untie or cut this knot; and whether it could be safely cut through, even by this all-powerful hand, remains questionable. It may not be at present a question of much interest to the British public, because no such large decisive experiment as was proposed has yet been tried, as to the value and attainableness of the object; but its magnitude and importance are incontestable. The whole extent of peat soil in Ireland exceeding, as "it is "confidently* pronounced, two millions eight "hundred and thirty thousand acres;" of which above half might be converted to the general purposes of agriculture.

Whenever the public attention shall become intent upon the subject, which must sooner or later happen, these "Reports" will be anxiously consulted; and it will, I trust, be found, that my father's contains, in as few words as possible, many applicable facts, some accurate experiments, some improvements on machinery now in use, and some new and ingenious mechanical contrivances.

In the year 1811, my father was occupied in constructing, upon a plan of his own invention, a spire for the church of Edgeworth-Town. This spire was formed of a skeleton

* Fourth and last Report of Commissioners, &c.

of iron, covered with slates, painted and sanded, to resemble Portland stone. It was put together on the ground, within the tower of the church, and when finished, it was to be drawn up at once, with the assistance of counterbalancing weights, to the top of the tower, and there to be fixed in its place.

The novelty of the construction of this spire, even in this its first skeleton state, excited attention; and as it drew toward its completion, and near the moment when, with its covering of slates, altogether amounting to many tons weight, it was to move, or not to move, fifty feet from the ground to the top of the tower, every body in the neighbourhood, forming different opinions of the probability of its success or failure, became interested in the event.

Several of my father's friends and acquaintance, in our own and from adjoining counties, came to see it drawn up. Fortunately it happened to be a very fine autumnal day—and the groups of spectators of different ranks and ages, assembled and waiting in silent expectation, gave a picturesque effect to the whole. A bugle sounded, as the signal for ascent. The top of the spire, appearing through the tower of the church, began to move upwards: its gilt ball and arrow glittered in the sun, while with motion that was scarcely perceptible it rose majestically. Not one word or interjection was uttered by any of the men, who worked the windlasses at the top of the tower.

It reached its destined station in eighteen minutes, and then a flag streamed from its summit, and gave notice, that all was safe. Not the slightest accident or difficulty occurred. The conduct of the whole had been trusted to my brother William, (the civil engineer); and the first words my father said, when he was congratulated upon the success of the work, were, that his son's steadiness in conducting business and commanding men gave him infinitely more satisfaction, than he could feel from the success of any invention of his own.

The spire was well secured, and provided with a conductor. This proved a wise precaution, for that very evening the weather changed, and a storm came on suddenly, with wind, thunder, and lightning. The storm lasted during the night, and in the morning the first thing of which we thought, and the first point to which we looked, was the spire: but my father himself was free from any anxiety on the subject.

Experience has hitherto justified his confidence; in seven years since its erection no change has been perceptible in the spire.

By a singular coincidence, the spindle of its weathercock was found to be precisely in the plane of the vertical wire of a transit instrument, in an observatory belonging to my father's house; and therefore, the slightest alteration of perpendicularity in the spire would be instantly detected.

An account of the construction of this spire, and of the simple machinery employed in raising it, may be found in Nicholson's Journal, vol. xxx, 1811*.

Towards the close of this year, 1811, Sir John Sinclair applied to my father, as he understood, by the desire of the Committee of the House of Commons, requesting that he would look over and give a report upon a mass of papers, containing the evidence and opinions of all, who had been examined by what has been technically called the *Committee of Broad Wheels.* The chief object was to determine, whether cylindrical or conical wheels are most advantageous, and to decide which are least destructive to the roads. To arrange, analyse, and report on all the heterogeneous evidence and opinions of carters, waggoners, coachmen, practical

* In the same Journal, vol. xxxi, 1812, is another paper relative to the same subject.

mechanics, and mechanic philosophers, from the civil engineers of the highest reputation, down to the most illiterate English drayman, was a tedious and difficult task. It was, however, after some months' labor, accomplished, with the assistance of his young secretary, my sister Honora, whose indefatigable patience and cheerful sympathy, as he ever felt pleasure in recollecting, made what would have been irksome delightful. After winnowing the sense from the nonsense, he gave the information, compressed into the smallest compass, and in the most convenient form. In the form of an alphabetical index, a *catalogue raisonné* was made, presenting the result and value of the evidence of each person, who had been examined.

When this was effected, and when he had prepared a report to be laid before the House of Commons, it was explained to him, that Sir John Sinclair had not been empowered to make his application by the Committee, of which he had been chairman. It was a private request of his own, and Sir John now expressed a wish, to have the report for the Board of Agriculture. With this request my father declined to comply; but to shew his respect for the members of that Board, and his willingness to contribute towards their useful and patriotic purposes, he wrote and presented to them "*An Essay on Springs applied to Carts.*"

He reserved for future publication what he had prepared as a report, resolving to add to it, from time to time, at his leisure, whatever further knowledge fresh observation and experiment might furnish. The Board of Agriculture did him the honor to vote him their thanks, and a hundred guineas to be presented to him, in the form of a piece of plate, or in any other form he might wish; but this he declined accepting, at the same time expressing a wish, that, if it should suit the views of the Society, the money might in future be applied towards defraying the expense of some public experiments on this subject.

Two years afterwards, when he was in London, he pursued this idea, and endeavoured to have a course of experiments tried on a large scale, to determine the draught of different carriages. The Duke of Bedford, to whom he had been known, when his Grace was Lord Lieutenant of Ireland, offered his assistance, with all the means of accommodation and publicity, which Woburn could so amply afford. But unfortunately some private affairs just at that moment occurred, which called my father away from London, and deprived him of the possibility of availing himself of these advantages. It was a disappointment, which he then and afterwards strongly lamented. He ever retained a grateful sense of the peculiar politeness shewn to him on this occasion by the Duke of Bedford, and felt true esteem for the patriotic spirit and zeal so worthy of a British nobleman, which His Grace has always shewn, to promote the interests of science, and to urge objects of national utility.

Before he left England (in 1813) my father published his essay on Roads and Wheel Carriages, in which he gave to the public all the information which he had collected, and fully explained the means, by which experiments on a large scale might be satisfactorily tried, to decide practically the theoretic points yet remaining in dispute.

To the first edition of this work was added, in the form of an appendix, the summary which he made of the information given by all who had been examined by the committee on Broad Wheels: but notwithstanding all the power of compression which had been exerted, this appendix proved too bulky and heavy for the generality of readers. In a second edition, the book was freed from what had been found an incumbrance.

I cannot leave this subject without observing, that, in the course of the drudgery which he went through, he received a great counterbalancing pleasure from the following passage, which he chanced to meet with in a letter to the committee written by a gentleman, to whom he was personally a stranger.

Mr. Edgeworth (Richard Lovell Edgeworth,) " was the " first, who pointed out the great benefit of springs in aiding " the draught of horses. The subject deserves more atten- " tion, than it has hitherto met with. No discovery relative to " carriages has been made in our time of equal importance; " and the ingenious author of it deserves highly some mark of " public gratitude.

" Yet so slow are improvements in getting into notice, that " though this fact was published (by him) more than twenty " years ago, it is still so little known, that the following erro- " neous opinion, by a gentleman who ought to have known " better, has been published, and is again repeated in the first " Report:—' *It is sufficiently obvious, that a short, stiff,* " *unelastic carriage, must be more easily drawn, than any* " *formed on a contrary principle.*' The known liberality of " this gentleman will certainly lead him to acknowledge his " error on this subject, as soon as he has perused Mr. Edge- " worth's excellent Memoir on Springs.

" John Whitley Boswell."

That memoir is in the Transactions of the Royal Irish Academy for the year 1788.

Since the time of Dr. Hooke it had been well understood, that springs contribute considerably to the ease, convenience, and safety of all who travel in carriages, or of goods conveyed in them; but no one had discovered, or at least had ever hinted to the public, the great utility of springs, in easing to the horses the draught of carriages, till this theory of my father's was given, explained, and supported by experiment in his memoir. It was stated in so few words, and with so little ostentation of discovery, that many perhaps read it, without knowing that it was new; or they passed it over without conceiving the important consequences, to which it would lead. As Mr. Boswell observes, it lay long unheeded as theory, by the many who call themselves practical men; by others more wise and more crafty, it had been used, appropriated, and amplified on, without due acknowledgement in any scientific publication. But it happened, that, in 1803, on the very day when my father, for the first time, arrived in Edinburgh, Professor Robison was lecturing on this branch of mechanics; and, accidentally going to the lecture, he had the satisfaction of hearing his own experiments mentioned, in the most gratifying manner, by that excellent judge of the subject.

I hope it will be observed, that through life he pursued steadily the same objects of research, with a perseverance, which proves his earnestness to arrive at truth. For example, in mechanics, he began by turning his attention to wheel-carriages: the first Essay he ever published was a letter on this subject, printed in the Museum Rusticum, when he was about nineteen; through all the years of his acquaintance with Dr. Darwin and Sir Francis Delaval, he continued to pursue the same subject; and at five-and-thirty, and at seventy, we find him prosecuting these inquiries, and still later in life we shall see him persevering in the same course.

Those ingenious ideas, which had been but the amusement of youth, as he advanced in life, he turned to public utility; for instance, the mode of conveying secret and swift intelligence, which he had suggested at first only to decide a trifling wager between him and some young nobleman, he afterwards improved into a national telegraph, and, through all difficulties and disappointments, persevered till it was established. In the same manner, his juvenile amusements with the sailing chariot led to experiments on the resistance of the air, which in more mature years he pursued in the patient spirit of philosophical investigation, and turned to good account for the real business of life, and for the advancement of science.

On this subject, in the year 1783 he published in the Transactions of the Royal Society, (vol. 73,) "An Essay on the Resistance of the Air," of which the object, as he states, is to determine the force of the wind upon surfaces of different size and figure, or upon the same surface, when placed in various directions, inclined at different angles, or curved in different arches. These experiments were originally suggested by a singular circumstance, mentioned by Mr. Robins, in his Treatise on Gunnery: "If two similar cards," says he, "are placed opposite the wind, one upon its end, and the other on its side, and both inclined to the horizon at the same angle, the wind will have the greater effect upon the card that is placed endways." To verify this experiment, and to pursue the inquiry, my father contrived an apparatus of a more accurate construction, upon a larger scale, and less liable to friction, than that which had been used by Mr. Robins. After trying several experiments on surfaces of various shapes, he ascertained the difference of resistance in different cases; suggested the probable cause of these variations; and opened a large field for future curious and useful speculation; *useful* it may be called, as well as curious, because such knowledge applies immediately to the wants and active business of life, to the construction of wind and water-mills, and to the extensive purposes of navigation. The theory of philosophers, and the practice of mechanics and seamen, were, and perhaps are still at variance as to the manner in which sails of windmills, and of ships, should be set. Dr. Hooke, in his day, expressed "his surprise, at the obstinacy of seamen, in continuing, after what appeared the clearest demonstration to the contrary, to prefer what are called bellying or bunting sails, to such as are hauled tight." The doctor said, that he would, at some future time, add the test of experiment to mathematical investigation in support of his own theory.

It is remarkable, that this test of experiment, when at length it was applied, confirmed the truth of what the philosopher had reprobated as an obstinate vulgar error. My father, in his "Essay on the Resistance of the Air," gives the result of his experiments on a flat and curved surface of the same dimensions, and explains the cause of the error, into which Dr. Hooke, M. Parent, and other mathematicians, had fallen in their theoretic reasonings.

Though at the hazard of making mistakes, even in copying and repeating, on a subject on which I am ignorant, I have ventured thus to state the general object and result of these experiments, in hopes of furthering my father's wish of turning public attention to the subject. Some, who might not have chanced to look back at the Philosophical Transactions of that time, may now, perhaps, feel their curiosity excited, and may recur to his essay for further information.

It is remarkable, that a man of naturally lively imagination, and of inventive genius, should not, in science, have ever followed any fanciful theory of his own, but that all he did should have been characterized by patient investigation, and prudent experiment.

It is tempting to human ingenuity, and still more tempting to vanity, to raise, each man after his own fashion, a little pyramid of human knowledge; but the first opposing wind or wave will destroy these, or they will perhaps be thrown down by his change of fancy.

> "The sportive *genius*, pleas'd with some new play,
> Sweeps the light works and fashion'd domes away."

The memory of his labors will be the more likely to endure, who has been content to work humbly after or along with others, even if he have added but one stone, one morsel of cement, to the great solid pyramid of human knowledge. In science, it is not given to man *to finish*: to persevere, to advance a step or two, is all that can be accomplished; and all that will be expected by the real philosopher.—"*We will endeavor,*" is the humble and becoming motto of our philosophical society *.

CHAPTER XVI.

IT has been said, that the best part of an author is always in his books. Upon this principle, the biography of literary men generally consists of little more than the history of the books which they have written, the dates of their publication, their different editions and variations, with an account of the quarrels and controversies, that may have occurred with brother authors, and a display of the encomiums of friends, or a repetition of complaints of the injustice of critics. All such tiresome topics the reader of these memoirs has been spared the more readily, because the best part of him, whose life I am writing, was *not* in his works. He undervalued his literary talents, so much, that for many years of his life he said, that he could not *write.* In one of his early letters to Mr. Day, the reader may recollect his concluding with

* Royal Irish Academy.

"You know I am no writer; my ideas do not, like yours, flow to my pen readily." He thought, that the early habit of writing Latin had given a Latin construction to his style; and I have often heard him declare, that he never wrote any thing he could bear to read till he was past forty. He never seemed to be satisfied with his own writing, when he knew it to be his. In general, he so completely forgot what he had written, that we have often read passages to him, without his recognizing them; and have cheated him into praising these, when he believed them to have been written by another. Though he had passed much of his youth with some of the most celebrated writers of his age, this did not excite in him any ambition to become an author, he was content with assisting and enjoying the celebrity of his friends.

One little book, however, he and Mrs. Honora Edgeworth began, I believe, in the year 1778, when she, in teaching her first child to read, found the want of something to follow Mrs. Barbauld's lessons, and felt the difficulty of explaining the language of the books for children, which were then in use.

> "Favete linguis——
> Virginibus puerisque canto,"

was the motto of this little volume, which was the first part of "Harry and Lucy," or of

"*Practical Education,*" as I find it called in the titlepage to the first copies, printed literally for his own children, and not published for many years afterwards. He intended to have carried on the history of Harry and Lucy, through every stage of childhood; to have diffused, through an interesting story, the first principles of morality, with some of the elements of science and literature, so as to shew parents how these may be taught, without wearying the pupil's attention. Much of this plan has, in different forms, been since executed in various publications, by persons of information and talents, who have, of late, written for young people; but at the time to which I refer, the design was new, and scarcely any English writer of eminence, except Dr. Watts and Mrs. Barbauld, had condescended to write for children.

Mr. Day, who was much pleased with my father's plan, offered to assist him, and with this intention began Sandford and Merton, which was first designed for a short story, to be inserted in Harry and Lucy. The illness of Mrs. Honora Edgeworth interrupted the progress of that little volume, and after her death, the ideas associated with it were so painful to my father, that it was not at that time continued. Meanwhile, Mr. Day wrote on rapidly, and finished and published his delightful book. Often, with pride and pleasure, my father used to say, that the public owed Sandford and Merton to him, since it was he, who first turned his friend's attention to the subject, and excited him to write an elementary work on education. After Harry and Lucy had remained, for above twenty years, with the first part printed, but not published; it was at last given to me, for a part of "*Early Lessons.*"

In fact, my father never exerted himself to write, or thought of becoming an author, till he felt sufficient motive, in the wish to encourage and assist me to finish "Practical Education." All his literary ambition then and ever was for me.

After "Practical Education," the next book which we published in partnership was (in 1803) the "Essay on Irish Bulls." The first design of this Essay was his:—under the semblance of attack, he wished to shew the English public the eloquence, wit, and talents of the lower classes of people in Ireland. Working zealously upon the ideas which he suggested, sometimes, what was spoken by him, was afterwards written by me; or when I wrote my first thoughts, they were corrected and improved by him; so that no book was ever written more completely in partnership.

On this, as on most subjects, whether light or serious, when we wrote together, it would now be difficult, almost impossible, to recollect, which thoughts originally were his, and which were mine. All passages, in which there are Latin quotations or classical allusions, must be his exclusively, because I am entirely ignorant of the learned languages. The notes on the Dublin shoe-black's metaphorical language, I recollect, are chiefly his.

I have heard him tell that story with all the natural, indescribable Irish tones and gestures, of which written language can give but a faint idea. He excelled in imitating the Irish because he never overstepped the modesty or the *assurance* of nature. He marked exquisitely the happy confidence, the shrewd wit of the people, without condescending to produce effect by caricature. He knew not only their comic talents, but their powers of pathos; and often when he had just heard from them some pathetic complaint, he has repeated it to me while the impression was fresh. In the chapter on wit and eloquence, in Irish Bulls, there is a speech of a poor freeholder to a candidate, who asked for his vote; this speech was made to my father, when he was canvassing the county of Longford. It was repeated to me a few hours afterwards, and I wrote it down instantly, without, I believe, the variation of a word.

In the same chapter there is the complaint of a poor widow against her landlord, and the landlord's reply in his own defence. This passage was quoted, I am told, by Campbell, in one of his celebrated lectures on eloquence. It was supposed by him to have been a quotation from a fictitious narrative, but, on the contrary, it is an unembellished fact. My father was the magistrate, before whom the widow and her landlord appeared, and made that complaint and defence, which he repeated, and I may say acted, for me. The speeches I instantly wrote word for word, and the whole was described exactly from the life of his representation.

After the "Essay on Bulls," my father never published any thing for several years, except some elementary books, suggested by feeling the want of them in his own family. In 1802 he published "Poetry explained for Young People": in 1816, "Readings on Poetry": and at various times, different parts of "Early Lessons." He also explained and illustrated his method of teaching to read, in a small tract, called "A Rational Primer." No man, who knew the world as well as he did, would have put his name to such books, or would have condescended to devote his time and talents to writing them, had his object been literary celebrity. Nothing but the true desire to be useful could have induced any man of talents, to choose such inglorious labors; but he thought no labor, however humble, beneath him, if it promised improvement in education. To the construction of twenty pages of a "Rational Primer" he devoted more time, than it would have cost him to write an octavo volume on another subject. It gave him more trouble, than those who are not used to the difficulties of early instruction can be aware, that the subject demands, or deserves.

In the introduction to this primer, a striking fact is asserted relative to the short time in which a child, taught by the method there described, learned to read. I have to add, that further trials upon different children in his own family encreased his confidence in the success of his plan. The quickness, with which the pupil may learn, he considered but as an inferior and comparatively unimportant object. His purpose was, to prevent children's associating with their first ideas of literature the painful feelings of overstrained or disappointed attention; and it was in this point of view chiefly, that he considered his mode of teaching to read as of material consequence.

His principle of always giving distinct marks for each different sound of the vowels has been since brought into more general use. It forms the foundation of Pestalozzi's plan of teaching to read. But one of the most useful of the marks in the "Rational Primer," the mark of obliteration, designed to shew what letters are to be omitted in pronouncing words, has not, I believe, been adopted by any public instructor. As to the method in general, more time than has yet elapsed must put it to the only sufficient trial, to that of experience, before we can decide on its expediency.

With respect to his other elementary books, "Readings on Poetry," and what he wrote of "Early Lessons," I have only to say, that children and parents have given him the satisfaction, which he most desired—the pleasure of knowing, that these books have answered the purpose, for which they were intended.

In 1808 he published "Professional Education." This work has now been a sufficient time before the public, to have been appreciated with that justice, which Time alone can grant. The best criticism and analysis that I have seen of this book, and of my father's general principles of Education, are in Rees's Encyclopedia, in the article, "Moral and Intellectual Education."

Compared with his powers of mind, my father wrote but little, but I may be permitted to say how much as a critic he did for me. Yet, indeed, this is out of my power, fully to state to the public—only that small circle of our friends, who saw the manuscripts before and after they were corrected by him, can know or imagine how much they were improved by his critical taste and judgment. Of the pains, the care he took, I may, perhaps, give some idea, by simply stating the facts.

The reader may, perhaps, recollect in one of his first letters to Mrs. Edgeworth in 1783, hearing of a translation of Adele and Theodore, which he was correcting, and which he sat up a night to finish. Another translation appeared, just as this was completed, and his labor in this revision was so far lost; but he was quite satisfied, because, as he said, he had attained his chief object, which was by example, as well as by precept, to excite and enure his pupil to application. I was at that time almost a child, and to ensure my perseverance, he had made me promise, that, if I began, I would finish; this would never have been accomplished, but for his sympathy, the interest he shewed in my progress, and the large share of labor he took himself, which could not but excite a grateful emulation. Mr. Day, who had a horror of female authorship, was alarmed and shocked by my father's having permitted his daughter even to translate. Some examples of want of discretion, and of ill conduct, which he had seen in women of literary talents, had prejudiced him to such a degree against female literature, that at one time, he was nearly of Sir Anthony Absolute's opinion, that the extent of a woman's erudition should consist in her knowing her simple letters, without their mischievous combinations. He often repeated the lines, which, it was said, Dr. Johnson once quoted to a celebrated authoress:

> " Nor make to dangerous wit a vain pretence,
> " But wisely rest content with sober sense;
> " For wit, like wine, intoxicates the brain,
> " Too strong for feeble woman to sustain;
> " Of those that claim it, more than half have none,
> " And half of those that have it, are undone."

Mr. Day wrote a congratulatory letter to my father, when the publication of the translation of Adele and Theodore was prevented. This letter contained an eloquent philippic against female authorship, to which my father replied, defending the cause of female literature.

From their containing personal allusions, these letters were, I suppose, destroyed with those, which my father committed to the flames, for I have not been able to find them. The impression however, which the eloquence of Mr. Day's letter made, though I heard it read only once, at the time it was received, remained for years in my mind; and it was from the recollection of his arguments, and of my father's reply, that "Letters for Literary Ladies" were written nearly ten years afterwards. They were not published, nor was any thing of ours published, till some time after Mr. Day's death. Though sensible, that there was much prejudice mixed with his reasons; yet, deference for his friend's judgment prevailed with my father, and made him dread for his daughter the name of authoress. Yet though publication was out of our thoughts, as subjects occurred, many essays and tales were written for private amusement.

Among others written many years ago, was one called "the History of the Freeman Family." In 1787, my father, to amuse Mrs. Elizabeth Edgeworth, when she was recovering after the birth of one of my brothers, related to us every evening, when we assembled in her room, part of this story, which I believe he invented as he went on. It was found so interesting by his audience, that they regretted much that it should not be preserved, and I in consequence began to write it from memory. "The plan, founded on the story of two families, one making their way in the world by independent efforts, the other by mean arts, and by courting the great, was long afterwards the groundwork of "Patronage." The character of Lord Oldborough was added, but most of the others remained as my father originally described them: his hero and heroine were in greater difficulties than mine, more in love, and consequently more interesting, and the whole story was infinitely more entertaining. I mention this, because some critics took it for granted, that he wrote parts of "Patronage," of which, in truth, he did not write, to the best of my recollection, any single passage; and it is remarkable, that they have ascribed to him all those faults, which were exclusively mine; the original design, which was really his, and which I altered, had all that merit of lively action and interest, in which mine has been found deficient.

Whenever I thought of writing any thing, I always told him my first rough plans; and always, with the instinct of a good critic, he used to fix immediately upon that, which would best answer the purpose.—"*Sketch that, and shew it to me*"—These words, from the experience of his sagacity, never failed to inspire me with hope of success. It was then sketched. Sometimes, when I was fond of a particular part, I used to dilate on it in the sketch; but to this he always objected —"I don't want any of your painting—none of your drapery!—I can imagine all that—let me see the bare skeleton."

It seemed to me sometimes impossible, that he could understand the very slight sketches I made; when, before I was conscious that I had expressed this doubt in my countenance, he always saw it.

"Now my dear little daughter, I know, does not believe, that I understand her."—Then he would in his own words fill up my sketch, paint the description, or represent the character intended, with such life, that I was quite convinced he not only seized the ideas, but that he saw, with the prophetic eye of taste, the utmost that could be made of them. After a sketch had his approbation, he would not see the filling it up, till it had been worked upon for a week or fortnight, or till the first thirty or forty pages were written; then they were read to him, and if he thought them going on tolerably well, the pleasure in his eyes, the approving sound of his voice, even without the praise he so warmly bestowed, were sufficient and delightful excitements to "go on and finish." When he thought that there was spirit in what was written, but that it required, as it often did, great correction, he would say, "Leave that to me: it is my business to *cut* and correct—yours to write on." His skill in *cutting*—his decision in criticism was peculiarly useful to me. His ready invention and infinite resource, when I had run myself into difficulties or absurdities, never failed to extricate me at my utmost need. It was the happy experience of this, and my consequent reliance on his ability, decision, and perfect truth, that relieved me from the vacillation and anxiety to which I was so much subject, that I am sure I should not have written or finished any thing without his support. He inspired in my mind a degree of hope and confidence, essential in the first instance to the full exertion of the mental powers, and necessary to ensure perseverance in any occupation. Such, happily for me, was his power over my mind, that no one thing I ever began to write was ever left unfinished.

Independently of all the advantages, which I as an individual received from my father's constant course of literary instruction, this was of considerable utility in another and less selfish point of view. My father called upon all the family to hear and judge of all we were writing. The taste for literature, and for judging of literary composition, was by this means formed and exercised in a large family, including a succession of nine or ten children, who grew up during the course of these twenty-five years. Stories of children exercised the judgment of children, and so on in proportion to their respective ages, all giving their opinions, and trying their powers of criticism fearlessly and freely. The sympathy with numbers, the mixture of the younger with the elder parts of the family in one and the same literary interest, was, in every point of view, advantageous. Every individual, feeling for or with the author, found his attention excited and kept up in discussing points of criticism, which might otherwise have been tiresome. My father listened with such acuteness of attention and affectionate avidity, that not a word escaped him—I do not say, not a *fault*.

My father's sympathy, in whatever I wrote, went through or along with every other interest of his mind: with the most kind consideration and address, he always managed so, that the reading of any thing I had prepared should be at the most agreeable times not only for himself, but for the whole family.

When he had been tired with the morning's business, he used in the evening to call upon me, to read something to refresh and entertain him.

He would sometimes advise me to lay by what was done for several months, and turn my mind to something else, that we might look back at it afterwards with fresh eyes. On the advantages of this practice, in confirmation of Horace's old precept, he pointed out to me some observations of Dr. Johnson's*, which are so just in thought, and forcible in language, that they made an indelible impression on my mind.

Many things I had written lay by several years, while I was occupied on others; and they were reconsidered by my father, read again at long intervals, and recorrected with such drudgery of revision, as nothing but the strength of affection could have made supportable to a man of his vivacity and his genius.

Were it worth while, I could easily point out many hints for invention, furnished by the incidents and characters he had met with in his youth, and which he related to me. But the reader will be best pleased with discovering these for himself in the preceding memoirs.

---

* Rambler, No. 169.

I may mention, because it leads to a general principle of criticism, that, in many cases, the attempt to join truth and fiction did not succeed: for instance, Mr. Day's educating Sabrina for his wife suggested the story of Virginia and Clarence Hervey in Belinda. But to avoid representing the real character of Mr. Day, which I did not think it right to draw, I used the incident, with the fictitious characters, which I made as unlike the real persons as I possibly could. My father observed to me afterwards, that, in this and other instances, the very *circumstances*, that were taken from real life, are those that have been objected to as improbable or impossible; for this, as he shewed me, there are good and sufficient reasons. In the first place, anxiety to avoid drawing the *characters*, that were to be blameable or ridiculous, from any individuals in real life, led me to apply whatever *circumstances* were taken from reality to characters quite different from those to whom the facts had occurred; and, consequently, when so applied, they were unsuitable and improbable: besides, as my father remarked, the circumstances, which in real life fix the attention, because they are out of the common course of events, are for this very reason unfit for the moral purposes, as well as for the dramatic effect of fiction. The interest we take in hearing an uncommon fact often depends on our belief in its truth. Introduce it into fiction, and this interest ceases, the reader stops to question the truth or probability of the narrative, the illusion and the dramatic effect are destroyed; and as to the moral, no safe conclusion for conduct can be drawn from any circumstances, which have not frequently happened, and which are not likely often to recur. In proportion as events are extraordinary, they are useless or unsafe, as foundations for prudential reasoning.

Besides all this, there are usually some small concurrent circumstances connected with extraordinary facts, which we like and admit as evidences of the truth, but which the rules of composition and taste forbid the introducing into fiction; so that the writer is reduced to the difficulty either of omitting the evidence on which the belief of reality rests, or of introducing what may be contrary to good taste, incongruous, out of proportion to the rest of the story, delaying its progress, or destructive of its unity. In short, it is dangerous to put a patch of truth into a fiction, for the truth is too strong for the fiction, and on all sides pulls it asunder.

Invention, it is said, is often overawed by criticism, and many writers have complained, perhaps with justice, of critics who can never suggest any thing new, in the place of that to which they object. Mine was a critic of a different sort; one who knew well both the difficulties and pleasures of invention—one, who, if he objected, knew how to remedy—who, even in assisting, knew how to give the writer all the pleasure of original composition. He left me always at full liberty to use or reject his hints, throwing new materials before me continually, with the profusion of genius and of affection. There was no danger of offending, or of disappointing him by not using what he offered. There was no vanity, no selfishness, to be managed with delicacy and deference; he had too much resource ever to adhere tenaciously to any one idea or invention. So far from it, he forgot his gifts almost as soon as he had made them—thought the ideas were mine, if they appeared before him in any form in which he liked them; and if never used, he never missed, never thought of inquiring for them. Continually he supplied new observations on every passing occurrence, and wakened the attention with anecdotes of the living or the dead. His knowledge of the world, and all that he had had opportunities of seeing behind the scenes in the drama of life, proved of inestimable service to me; all that I could not otherwise have known, was thus supplied in the best possible manner. Few female authors, perhaps none, have ever enjoyed such advantages, in a critic, friend, and father, united. Few have ever been blest in their own family with such able assistance, such powerful motive, such constant sympathy.

Forgive me, reader, for dwelling on those happy days; I could not indeed write my father's life justly, considering him either as a man of letters, or as a preceptor, if I had omitted these things.

CHAPTER XVII.

I HAVE not been able to carry on my father's biography during these latter years by any of those private letters, which afford so full a view of the habits, occupations, and opinions, of the writer. After the death of Dr. Darwin, the last of the intimate correspondents of his youth, a new race of friends sprang up, and his letters to them are superior to his earlier letters in literary value, and characteristic strokes; but the persons to whom they are addressed are still living, and the mixture of subjects renders their publication impossible.

As I always lived with my father, I have but very few of his letters; but fortunately I have one or two, which may further shew how we went on together as literary partners.

TO MISS EDGEWORTH.

"MY DEAR MARIA, *August* 4, 1804.

"Your critic, partner, father, friend, has finished your Leonora. He has cut out a few pages; one or two letters are nearly untouched; the rest are cut, scrawled, and interlined without mercy.

"I make no doubt of the success of the book, amongst *a certain class of readers*, PROVIDED it be reduced to one small volume, and provided it be polished *ad unguem*, so that neither flaw nor seam can be perceived by the utmost critical acumen.

"As it has no story, to interest the curiosity; no comic, to make the reader laugh; nor tragic, to make him cry—it must depend upon the development of sentiment, the verisimilitude of character, and the elegance of style, which the higher classes of the literary world expect in such a performance, and may accept, in lieu of fable, and of excitement for their feelings. These you well know how to give; and your honest gratitude towards a favoring public will induce your accustomed industry, to put the highest finish to the work. For this purpose I advise you, to revise it frequently, and look upon it as a promising infant committed to your care, which you are bound by many ties to educate, and bring out when it is fit to be presented. The design is worthy of that encouragement, which you have already received; it rests on nature, truth, sound morality, and religion; and if you polish it, it will sparkle in the regions of moral fashion.

"You will be surprised to hear, that I have corrected more faults of style in this, than in any thing I have ever corrected for you. Your uncle Ruxton's criticisms have, except one, been adopted by me; and I hope when you have corrected it again, that he will have the goodness to revise it a second time.

"Give my love to little F * * * *, and tell her, that I had not time to explain a section to her. I therefore beg, that, with as little explanation as possible, you will bisect a lemon before her, and point out the appearance of the rind, of the cavities, and seeds; and afterwards at your leisure get a small cylinder of wood turned for her, and cut it into a transverse section, and into a longitudinal section.

"R. L. E."

My father was averse to carrying on any regular correspondence, writing to his friends only when prompted by some immediate subject, either of business or affection. His most regular correspondence was with the late excellent Joseph Johnson, his bookseller—the man of whom the poet Cowper speaks so frequently in his letters with strong regard, and with expressions of just esteem. All, who have ever had any connexion with him, think of him, I believe, in the same manner. I can answer for my father's having had the highest opinion of his professional integrity, of his humane, liberal, benevolent character, and of his amiable temper.

He considered Mr. Johnson quite as a friend. Indeed he had reason to do so. It is difficult, without hurting the feelings, or in some way compromising the characters of others, to bring before the public examples in support of these assertions; but, suppressing names and dates, one fact I may mention.

A manuscript, written by a celebrated person, and containing the life of one still more interesting to the public, was offered to Mr. Johnson. He knew, that it was likely to be a popular work, and, on the terms on which it was offered, it would have been to him a very advantageous speculation: but, it contained some abuse of my father (of what nature I do not know). Mr. Johnson returned the manuscript, refusing to have any thing to do with it, unless these passages were struck out. The invective was consequently obliterated, and some flattery substituted in its stead. Not till a length of time afterwards, and then only through the indiscretion of one who knew the secret, did it transpire to one of our friends, who thought it would give my father pleasure, to hear of our bookseller's handsome conduct.

The last letter poor Johnson ever wrote, or, I should say, ever dictated, was to my father. It was in his nephew's hand, and communicated to us the following account of his death.

London, 21*st December*, 1809.

DEAR SIR,

I am sorry it falls to my lot, to communicate to you the melancholy and sudden change, which has taken place in our family. My ever to be lamented uncle, Mr. Johnson, terminated his honourable life yesterday afternoon, about four o'clock. His illness was short, for, on Sunday morning, he was as well as usual; since which time, an inflammation in the lungs took place, and baffled all the exertions of his medical attendants. A short time before he died, he dictated the following words, and soon after expired.

" My uncle is so afflicted with the spasms and asthma, that " he has desired me to write to you, to say, that he should ill " deserve your confidence, if he were rigidly to adhere to the " contract, which he made for the last work; the sale of " which has enabled him, to double the original purchase money, " and to place the sum to the credit of your account."

I remain,

With the greatest respect,

Your most obedient Servant,

JOHN MILES.

Mr. Miles was alone with Mr. Johnson, when this letter, this dying order was dictated, which involved a considerable sum. What Mr. Johnson so generously ordered was immediately paid by his executors in the most handsome manner, with an alacrity, which shewed that his representatives felt pleasure in maintaining his character for liberality.

Some time after Mr. Johnson's death, an engraved portrait of him was published, of which his nephews sent us a copy. My father wrote under the print the following lines.

Wretches there are, their lucky stars who bless
Whene'er they find a Genius in distress:
Who starve the Bard, and stunt his growing fame,
Lest they should pay the value for his name.
But JOHNSON raised the drooping Bard from earth,
And fostered rising Genius from its birth:
His liberal spirit a *profession* made
Of what, with vulgar souls, is vulgar *trade*.

R. L. E.

Two years after Mr. Johnson's death, we had opportunities of seeing more of the liberal character of his successors, in their conduct to a very deserving person, whom we introduced to them, Mrs. Leadbeater, author of "Cottage Dialogues." She trusted entirely to them, and had ample reason to be satisfied. My father, who was as much pleased as she could be, wrote warm at the moment the following note.

"MY DEAR GENTLEMEN,

"I have just heard your letter to Mrs. Leadbeater read, by one who dropped tears of pleasure, from a sense of your generous and handsome conduct. I take great pleasure in speaking of you to the rest of the world, as you deserve; and I cannot refrain from expressing to yourselves the genuine esteem, that I feel for you. I know, that this direct praise is scarcely allowable, but my advanced age, and my close connexion with you, must be my excuse.

"Yours, sincerely,

"*Edgeworth-Town, May the* 31*st*, 1811. "R. L. E.

"*My* 68*th Birth-day*."

"*Collon,* 1*st September,* 1813.

"MY DEAR DAUGHTER, PARTNER, AND FRIEND,

"I have just read 'Harry and Lucy,' to my great satisfaction, and to my great mortification. I am mortified, by finding from your mother, that it is put into the hands of children of four years old, for whom it is totally unfit; and of course, they do not like it when they first read it, and this association of *displeasure* prevents them from liking it afterwards, when it might be serviceable.

"On the other hand, I felt satisfaction, from finding, that it contained the seed and roots of much useful fundamental knowledge, which is clearly explained, and adroitly interwoven with the story.

"Your 'Frank' should precede Harry and Lucy, in the series of Early Lessons; and there should be a preface, in which the order in which they should be read should be pointed out; and mothers should be strongly urged to postpone the use of the second part of Harry and Lucy, till their children should be eight years old at the least.

"It appears to me, that books for very young readers should consist of short sentences, or names of things; and that continued stories, and all attempts at instructive dialogue or narrative, should be delayed, till the child has learned to read a large number of monosyllables perfectly at sight. I well remember the first sentence, which I ever understood:—'*dogs bark at a noise.*'

"I must enquire from our bookseller, what books for very young children are at present in circulation; indeed, the best mode of enquiry is, to desire him to send all that he thinks tolerable.

"Names of things, with *good* prints, seem to me, to be the best adapted of any thing I can devise, for teaching the rudiments of reading. This has been practised, both in England and in France; but the prints have been so indifferently executed, as to hurt the picturesque eye of the child; if the prints were good, the pictured forms of things would be early associated with their names. We may all observe, in our own minds, what barbarous figures occur to us, when we think in the abstract of a horse, or a rose, or a candlestick. This may arise from the bad representation in prints, which we first saw of these things. Distinguishing a *B* from a *bull's foot* is sometimes a proof of more intellect, than is supposed to be necessary for such discrimination.

" I perceive, that Harry and Lucy has more merit than I had supposed; and what is really singular, (or rather, two-fold), I do not know, whether the latter part of it was written by you, or by me. I shall go on with it at my leisure.

" F——" (a child of four years old), " shews, every moment, the fruits of his careful education. He sees more, and understands more of what he sees, than any child I have ever met with. Mills, houses, roads, looms, whatever he examines, add not only to his stock of words, but to his ideas; and by having acquired the habit of speaking accurately, and of listening to the names of things, his technical vocabulary is always adequate to the explanation of his thoughts.

" I was really surprised at the facility, with which he understood the process of weaving, by passing threads crossways between longitudinal threads; of keeping the warp stretched by the beam; of the use of the reed; of the shuttle, and of the manner in which the shuttle is made to run through the warp, by its being struck by the little *battledores*, which the weaver moves by a handle and a string.

" His conduct is, in every particular, such as should be wished, and he has not been of the least trouble to any body, since he left home.

" R. L. E."

It is remarkable, that none even of these my father's most familiar letters, nor any thing in his memoirs, nor any thing he has published, can give the slightest idea of his powers and manner of conversation. His style in speaking and writing were as different, as it is possible to conceive. In writing, cool and careful, as if on his guard against his natural liveliness of imagination; he was so cautious to avoid exaggeration, that he sometimes repressed enthusiasm. The character of his writings, if I mistake not, is good sense; the characteristic of his conversation was genius and vivacity—one moment playing on the surface, the next diving to the bottom of the subject. When any thing touched his feelings, exciting either admiration or indignation, he poured forth enthusiastic eloquence; and then changed quickly to reasoning or wit. His transitions from one thought and feeling, or from one subject and tone to another, were so frequent and rapid, as to surprise, and sometimes to bewilder persons of slow intellect; but always to entertain and delight those of quick capacity. He had a constant flow of thought, joining with the current of other minds; thence, gathering fresh strength, not headlong in its course, but easily turning with every bend in its progress.

His observations pleased at the moment, perhaps beyond their just comparative or positive value, from the conviction felt of their coming fresh and full from the mind. We naturally believe, that he has great resources, who seems to set no store by the wealth he scatters.

During his intimacy with Mr. Day, he adopted, perhaps, too much of his friend's taste for long arguments; experience convinced him, that these protracted discussions seldom ended in any satisfactory conclusion, either to the understanding or to the temper. As he became fonder of writing, he found it more advantageous, to discuss on paper subjects that require precision: and he quite laid aside the set, single-combat style of conversation. But he always continued to require, that, when people did argue, they should reason with precision; and he was provoked, when any one substituted sentiment, assertion, authority, or opinion, instead of argument.

He used to say, half in jest, half in earnest, that he would rather a friend should give him a blow, than refuse him a reason. Though, from his vivacity and epigrammatic facility, it might have been expected, that wit would have pleased him more than reason, yet from those, for whom he was most interested, he shewed greater satisfaction in hearing a good reason, than any stroke of wit.

His openness in conversation went too far,

almost to imprudence; exposing him not only to be misrepresented, but to be misunderstood. Those, who did not know him intimately, often took literally what was either said in sport, or spoken with the intention of making a strong impression for some good purpose. Whenever he perceived in any of his friends, or in one of his children, an error of mind, or fault of character, dangerous to their happiness; or when he saw good opportunity of doing them service, by apposite and strong remark or eloquent appeal in conversation, he pursued his object with all the boldness of truth, and with all the warmth of affection. Blaming or praising, it was the same; the same carelessness of what might be thought of himself; the same truth and feeling for others. His overflowing affection for his family, his confidence in the sympathy of those to whom he spoke, carried him sometimes beyond what indifferent auditors could follow, and he frankly laid himself open to the blame of all, who pride themselves upon that disqualifying sort of hypocrisy, which so commonly passes for modesty. Time, however, the constant friend to truth, settled his character on this as well as on many other points. Allowance was made for his habits of openness; and it was observed, that in what concerned *him-*

self alone, so far from overvaluing, he generally undervalued his talents.

Time gave him also in another way the full reward of his frankness. His character for sincerity and openness, his readiness to confide in his fellow-creatures expanded in return their hearts, and gave him in conversation that power, which neither high spirits nor brilliancy of talents, even when joined to a strong desire to please, can command; the power of inspiring freedom of thought and the happy feeling of confidence.

I will not deny what I have heard from some, whose truth and sense I cannot question, that his manner, somewhat unusual, of *drawing people out*, however kindly intended, often abashed the timid, and alarmed the cautious; but, in the judgments he formed of the understandings of all with whom he conversed, he was uncommonly indulgent. He allowed for the prejudices or for the deficiencies of education, and he foresaw, with the prophetic eye of benevolence, what the understanding or character might become, if certain improvements were effected. In discerning genius or abilities of any kind, his penetration was so quick and just, that it seemed as if he possessed some mental divining rod, revealing to him hidden veins of talent, and giving him the power of discovering mines of intellectual wealth, which lay unsuspected even by the possessor.

To young persons his manner was most kind and encouraging. I have been gratified by the assurance, that many have owed to the instruction and encouragement received from him in casual conversation their first hopes of themselves, their resolution to improve, and a happy change in the color and fortune of their future lives. In fact in many instances, the strong ambition he inspired, to deserve his approbation and esteem, became a permanent governing motive in the minds of persons even unconnected with him, or with his family.

His conversation, as he grew older, was still more agreeable, than what I can remember it to have been in his middle age, and far more pleasing in general society, than what it had been, as I have been told, in his youth. Time mellowed, but did not impair his vivacity; so that seeming less connected with high animal spirits, it acquired more the character of intellectual energy. Still in age, as in youth, he never needed the stimulus of convivial company, or of new auditors: his spirits and conversation were always more delightful in his own family, and in every-day life, than in company, even the most literary or distinguished.

It may be regretted, perhaps, that instead of a description of his conversation, I do not give some specimens. I have none; for I never made notes of what he said, except for the particular purpose of registering experiments in education, or of illustrating his method of teaching. And though I reproach myself for not having registered these more fully and frequently, yet I do not even now regret, that I have no other notes of his conversation. My father would have detested and despised any *worshipper*, who had followed him about to note down his sayings; and of all persons I was least suited to such work. It would have destroyed our mutual confidence, and would have been quite incompatible with the manner, in which we lived together. To the public it would have proved useless, or worse—fallacious; for though it would be easy to give false ideas of his mind and of his abilities, by scraps of his sayings; yet without producing a mass equal to that which Boswell collected of Johnson, it would be impossible to give to those, who did not know him intimately, an adequate and just idea of his character from his conversation.

CHAPTER XVIII.

[1813.

THAT "sordid content," which my father in his letter to Lord Selkirk deplored as the bane of Ireland, and that despairing indolence and recklessness of the future, which formerly characterised the lower classes of the people, have, within the last thirty or forty years, considerably diminished; hope, industry, laudable ambition, habits of forecast and providence, have appeared. At the time to which I now allude, that is just before the peace (in 1814), this was strikingly apparent in the increase of the comforts, as well as the necessaries of life among the poor. Speaking without exaggeration, I may say, that increasing *attempts* towards order and neatness in their dwellings, apparel, and mode of living, became in this part of Ireland, every where visible. Assertions are not so satisfactory as particular facts; and yet the statement of facts leads often to misconception, from the propensity of the English public to believe, that what is true of a part of Ireland is true of the whole; while on the other hand, it is not always possible, to enter into sufficient detail, to give information without fatiguing the fastidious taste of impatient, high-bred benevolence. Whoever will take the trouble to look at the reports of the Board of Education, and at the Statistical Surveys of the different counties of Ireland, will obtain much exact knowledge upon this subject; though in many of the county surveys this will be found loaded with superfluous, tedious matter. My father wrote a Statistical Survey of the County of Longford, as he mentions in his letter to Lord Selkirk, but it is not yet published. It waited so long for the finishing of a county map, that it will now be necessary, before it can be given to the public, to add to what he wrote a view of the changes, that have taken place in the county since 1806, when his survey was written. How much the whole county has been improved by large plantations, by buildings, and by better modes of agriculture, will then appear, and a judgment may thence be formed of the rate of general improvement.

My father regretted late in life, that he had not begun by planting his estate; he desired, that this regret should be mentioned for the advantage of others. Some calculations he made on the profits of planting are given in the Appendix.

The result of his long residence in Ireland

and of his attempts to improve the condition of the poor in his neighbourhood, shall now be stated.

The exertions he made from the time he settled at Edgeworth-Town in 1782, in building comfortable dwellings for some of his tenants, and in assisting others to build the same for themselves;—his never following the vile system of making forty shilling freeholders, merely for electioneering purposes—the reasonable rent and tenure at which he let his land—the unusual time which he allowed his tenants to *make* their rent—his freeing them from *duty work*—his avoiding as much as possible, in his leases, oppressive or restrictive clauses—his respecting the *tenant's right*, wherever tenants had improved—his encouraging them by the certainty of justice and kindness—his discouraging all expectation of partial favor or *protection*, if they transgressed the laws, or if they lived in indolence and inebriety;. succeeded altogether beyond his most sanguine hopes, in meliorating the condition of the people. Especially within the last twenty years, his tenantry, and the whole face of his estate, strikingly improved in appearance, and essentially in reality. The poorest class of his tenants, who in former times lived in smoke and dirt, in too pitiable a condition for description, have now to most of their cabins chimneys and windows, comfortable thatch, and good earthen floors. The dunghills no longer stop up the windows, nor is "the first step out of the cabin into the dirt." The number of slated houses and boarded floors has much encreased; and what is of more consequence, and of better promise, for the future permanence of good habits, and for the progress of improvement, much of what has been done has been effected, not by the landlord, but by the tenants. Even some of the poorest have exerted themselves, to make small additions and improvements in their habitations. No matter how small, my father always, from the first dawning of hope, hailed the appearance of these efforts with due encouragement and assistance; and this, more than any pecuniary donations, tended to increase the disposition to exertion. Even from the first mending of a gap, or the digging a rough cabbage-garden, the planting a hedge, or the building a rude wall, he had hopes; and in these he was not disappointed.

It may be proper to mention, for those who see things more in a prudential, than either in a sentimental or patriotic point of view, that his pecuniary interests did not ultimately suffer for any temporary sacrifices or forbearance, which he shewed towards his tenants. On the contrary, his land very considerably increased in value, from the effect of better cultivation; and this was proportionably felt in his income at the close of leases, when the land reverted to him. In other cases, where the leases did not terminate, but when the times altered at the peace, when the change of prices and the fall in demand ruined many, both landlords and tenants—and when the distress in this country was so great, that rents could not without difficulty be obtained—his were, with few exceptions, and with moderate allowance, regularly paid. As he advanced in years, my father had the satisfaction, and a very great delight it was to him, to see himself surrounded by a respectable, flourishing, independent, attached, grateful tenantry.

He was a friend, not only to his own tenantry, but to all within his influence as a country gentleman; not merely by relieving their temporary wants, but by protecting them as a magistrate from injustice and oppression, by instructing them as to their real interests, and shewing them the consequences of their bad habits. In this point of view also his residence in Ireland succeeded in doing good beyond his hopes. A number of human beings have been trained by him in the way they should go, and others, timely advised, have been saved from guilt and destruction. Reclaimed from the effects of ignorance and bad example, or raised from the indolence of despair, they have become good subjects, and useful members of society. In various lines of humble life, he educated and forwarded in the world many excellent servants, workmen, and tradespeople; and in classes much above these, several young persons, sons of tenants, who looked up to him for protection and advice, and whose early habits and principles he happily influenced, have advanced in different professions, and have succeeded in situations beyond his or their most sanguine expectations. He took pleasure and pride in counting the numbers of those, who from this remote little village have gone out into the world, and have made their way in foreign countries. Letters from many of these from France, Spain, America, from the East and West Indies—news of their success, evidence of their good conduct, and tokens of their affection and gratitude—have often, in his latter years, and to his latest days, gladdened my father's heart.

Without the permission of the writer—he is in India, and I cannot wait to obtain it—I will venture to publish two letters, which do honor to the writer's grateful heart, and which, as he will be pleased in feeling, prove, in one instance, the truth of what I have just said of my father.

FROM MR. W— B * * * *.

"DEAR SIR, 1805-6.

"The concern which always appeared in you for my welfare, and the many advantages which I have derived from that concern, are such as I can no otherwise than with the most heartfelt gratitude remember; indeed, such as would be out of the power of a more able linguist than I am to express.

"When I reflect on every circumstance of my acquaintance with you, particularly the commencement through my worthy friend and benefactor, Mr. Lovell (E.) in that light having seen something of the world enables me to do, I must say, that Providence favored me highly, by making an instrument of you to protect me from the worse effects of the precarious situation in which I was, and to be the means of leading me to that state, to which I have every hopes of succeeding. My acquaintance with you has not only been productive of happy consequence in a literary sense, but also in matters of perhaps as great importance, and which, until now, I have not seen so clearly into.

"The notice you took of me, and the good opinion which I saw you had formed of me, inspired me with a zeal to conduct myself with a degree of prudence, which is not common to young men in such a sphere as that in which I was placed; and again, my expectations of succeeding through you to a genteel situation in life kept me from taking any unpremeditated rash steps, or, I may say, from throwing myself away in a manner, which narrow circumstances are apt to cause young men of spirit to do: this last I plainly see would inevitably have been the case with me.

"So far I have said, to shew you how sensibly I feel the obligations, which I am under to you and your worthy family—I find myself no way equal to the task of doing my feelings on this head justice; but I know, that you would like to find gratitude in a person, whom you have condescended so far both to honor and befriend.

"The ship Hercules sailed from here to England the 24th of September last, by which conveyance my brother wrote you a letter, giving all the information of how matters went on then. This caused me to omit writing then, expecting, that by this opportunity I should have some accounts to give you of the Telegraph† * * * * Pray do me the honor to make my most grateful respects to your family; and when you write to Mr. Lovell, remind him of me; and as it is out of my power, to have the happiness of writing to him myself, pray assure him of the veneration and esteem, with which I do, and always shall remember him. I hope in God I shall yet have the—what shall I say?—happiness is too feeble—of seeing him once more. Nothing would be a greater pleasure to me, than to receive a letter from you, as I should expect from it a renewal of the edifying instructions and advice, which you so often honored me with, and from which I have imbibed so much good.

"I have the honor to be,

"Dear Sir,

"Your very grateful,

"Humble and obedient servant,

"W * * * * * * * * B * * * *."

---

† Mr. B * * * * had been employed by my father in the telegraphic establishment in Ireland. When he went out to India, he carried with him such a perfect recollection of the machinery, the vocabularies, and the management of the business, that he established a line of similar telegraphs in India. For part of this letter which gives an account of this undertaking, see Appendix.

The following was written some years after the preceding letter.

FROM W. B.

"*Bombay, 2d November,* 1814.

"WORTHY SIR,

"I never took up my pen to address you, with so much diffidence as upon the present occasion; but I trust, as usual, to your goodness.

"It has long been one of the greatest of my wishes, that your family should be in possession of something from me, as a small token of my remembrance, respect, and gratitude, in the event of any unforeseen accident occurring, to prevent my return to my native land, and deprive me of the happy opportunity of giving expression to those feelings personally; but, I find, that I cannot indulge this wish, without taking a liberty, which is not, perhaps, warranted by even the experience I have had of your condescension; I have ventured, however, to get an article made up for Mrs. Edgeworth, of which, I entreat you to solicit her acceptance, and which, I am sure, she will the more readily do, if you will be good enough to explain to her, that it is but of small value, and that I should not presume to ask the favor of her acceptance of it, were it not, that it is a curiosity which is not procurable *at home for money.* It is an Indian box, containing a lady's work-box and escrutoire, with apparatus complete: composed of sandal-wood, ivory, and ebony, and a composition of quicksilver and sea-shells, which resembles silver. The *whole* of the outer surface is a complete scale of inlaying of the above materials, dyed of different colours, and has a most beautiful appearance. It will be the greater curiosity at home, as there is but one family in this country, (a Mogul family,) who understands the art of making these boxes, and which has been handed down from father to son from time immemorial. From this cause also, but very few of them indeed can find their way to Europe, as it takes six months to complete one of them.

"Having presumed upon the liberty of begging Mrs. Edgeworth's acceptance of a token of my respect and remembrance, I could easily have found something of a more valuable kind, but nothing, that would, perhaps, be a greater curiosity at home; and being an article which is connected at once with industry and science, it is, I conceive, the greatest tribute I can pay to that lady * * * * * *

"I am, &c.

"W. B."

The writer of these letters is now in a good situation, in Bombay, with a highly respectable character.

The middle classes of gentry in this part of Ireland have, within these last thirty or forty years, improved much in their general mode of living, in manners, and in information. The whole style and tone of society are altered.—The fashion has passed away of those desperately tiresome, long, formal dinners, which were given two or three times a year by each family in the country to their neighbours, where the company had more than they could eat, and twenty times more than they should drink; where the gentlemen could talk only of claret, horses, or dogs; and the ladies, only of dress or scandal: so that in the long hours, when they were left to their own discretion, after having examined and appraised each other's finery, many an absent neighbour's character was torn to pieces, merely for want of something to say or to do in the stupid circle. But now, the dreadful circle is no more; the chairs, which formerly could only take that form, at which the firmest nerves must ever tremble, are allowed to stand, or turn in any way which may suit the convenience and pleasure of conversation. The gentlemen and ladies are not separated from the time dinner ends, till the midnight hour, when the carriages came to the door to carry off the bodies of the dead; or, till just sense enough being left, to find their way straight to the tea-table, the gentlemen could only swallow a hasty cup of cold coffee or stewed tea, and be carried off by their sleepy wives, happy if the power of reproach were lost in fatigue.

A taste for reading and literary conversation has been universally acquired and diffused. Literature has become, as my father long ago prophesied that it would become, fashionable; so that it is really necessary to all, who would appear to advantage, even in the society of their country neighbours. A new generation of well informed young people has grown up, some educated in England, some in Ireland; while those of former days have been obliged to change their tone of real or affected contempt for *reading people.* They have been compelled, either to cultivate themselves in haste, to keep pace with their neighbours, or to assume at least the appearance of understanding, and of liking that, which has become the mode.

About the year 1783 or 1784, my father happened to be present in the only great bookseller's shop then in Dublin, when a cargo of new books from London arrived, and among them, the Reviews, or *the Review,* for the Monthly Review was the only one then sufficiently in circulation, to make its way to Ireland. Of these, my father found on inquiry, that not above a dozen, or twenty at the utmost, were ordered in this island. I am informed, that more than two thousand Reviews are now taken in regularly. This may give some measure of the general increase of our taste for literature. The Edinburgh and Quarterly Reviews are now to be found in the houses of most of our principal farmers: and all therein contained, and the positive, comparative, and superlative merits and demerits of Scott, Campbell, and Lord Byron, are now as common table and tea-table talk here, as in any part of the United Empire.

The distinction, which about half a century ago was very strongly marked between the manners and mental cultivation of a few families of the highest class of the aristocracy in Ireland, and all of the secondary class of gentry, has now, by the diffusion of literature, and the general improvement in education, been softened so much, as to be effaced in its most striking points of contrast. What might be termed the monopoly of elegance and information, it is no longer possible to maintain; this may be mortifying in some few instances to pride, but good sense, to say nothing of benevolence or patriotism, will see ample compensation. Even to the few, who enjoyed preeminence, it must have been, if not painful, at least, attended with many inconveniencies and dangers: all the inconvenience of being misunderstood and misrepresented, and all the dangers of jealousy and envy. The pride of being without a rival is dreadfully overbalanced by the misfortune of being without a judge. The pleasure of sympathy and society, with free, equal, rational conversation, the sense of being justly appreciated, and the hope of obtaining the approbation of those, who are themselves worthy of praise, must in the scale of real happiness far outweigh all the joys of commanding awe, admiration, or fear, from the tribe of "stupid starers."

It is incalculably for the advantage of persons of talents, that knowledge and the taste for literature should be generally diffused; and this must be peculiarly felt by any individual, who has lived long enough in the same place and country, to contrast the difference made by these improvements, both on society in general, and on the estimation in which he himself was held. As society here became better informed, and more cultivated, my father was not only more valued and respected, but infinitely better liked. His disdain, and my wish to see him popular, formed a never failing subject of debate and raillery.

About this time he used to say, "Maria, I am growing dreadfully popular, I shall be good for nothing soon; a man cannot be good for any thing, who is very popular."

With this Coriolanus spirit his mother had early inspired him. But, notwithstanding what he might say to me in jest, he was seriously sensible both of the advantage and the pleasure of being justly appreciated. In consequence, society became more agreeable to him; cause and effect, in this instance, as usual, operating alternately and reciprocally.

In 1813, he visited England for the last time. During the London winter, or *natural* summer, which he spent there, he had the happiness of becoming acquainted with most of the celebrated persons of the present day, and with those most distinguished by their talents, their character, and their station in society. From his writings on education, and from the examples of his mode of domestic instruction, parents had become interested about him. Many of these of different classes, chiefly in the higher, I may say, the highest ranks, were desirous to converse with him, to obtain his counsels, and to hear his opinions on education. They would not allow him, to consider himself as a stranger to them. They spoke to him with grateful cordiality and frankness, as to a person long known and loved by their families.

Upon our return home in the autumn of that year, some of these did him the honour to visit him at Edgeworth-Town; and on seeing him in his own country, and in the midst of his family, their regard for him encreased with a rapidity not common to English habits. From acquaintance, they became, and ever afterwards continued friends—kind, intimate, warm friends to him and his.

He felt, with just pride and pleasure, the enjoyment of thus drawing round him in remote retirement such society, and of attracting persons of such superior merit; while we gratefully rejoiced, in seeing him thus possessed of all "that should accompany old age —honor, love, obedience, troops of friends."

And those who had known him longest, loved him best. The sister, who had been his earliest friend, the favorite companion of his childhood, and every surviving friend of his youth, we saw more strongly attached, and more dear to him in age. I have heard him say, that (with one exception) he never in the whole course of his life lost a friend, but by death. The happiest proofs of the truth of this assertion were in our knowledge, or before our eyes. We could even go farther from our own observation, and assert, that the esteem and affection of every person, whom he had ever called his friend, had not merely continued unabated, but had encreased as they had advanced in years, in proportion as they had greater opportunities of experience and comparison.

Fifteen years had now passed, since his last marriage. The sisters of a former wife continued to reside in his family, having become the most attached friends of the present Mrs. Edgeworth and of her children. Under her uniting influence, he saw his sons and daughters, by three previous marriages, living together with six of her children, all in perfect harmony and happiness; all looking up to him with fond affection, confidence, and gratitude. From the great difference in the ages of his children, his eldest being at that time above five and forty, the youngest, only one year old; he enjoyed as a father, preceptor, and friend, an extraordinary variety of interest and amusement, as well as occupation and friendship in his own family. Some had been for years his friends and companions, had joined with him in all his pursuits, thoughts, and feelings, and had lived with him on terms of equality, which, diminishing nothing from respect, added incalculably to our happiness, gratitude, and affection.

Then he had, both as a preceptor and a parent, continual interest in the education of his sons; while their gratitude, and the promise of their excellence, delighted their father's heart with the fairest prospect, and the most reasonable of human hopes. For the fortitude with which he had sustained former misfortunes, and the energy with which he had persevered unremittingly in the education of those which remained to him, he was rewarded by seeing a new race growing up round him, to supply, not to obliterate in his affections, those whom he had lost.

He was at this time in his seventieth year, but he had not as yet felt any of the infirmities of age; and as we, indeed as all who saw him thought, his activity of body, his vigour of intellect, his energy of character, his flow of spirits, and youth of heart, seemed to promise many years of life and health.

But early in the spring of 1814, and as it happened on the evening of a day, when a visitor, who saw him for the first time, had been particularly struck by his healthy appearance, and by that family happiness, of which he was the centre, and, under Providence, the cause, my father was suddenly seized by an alarming and painful illness *. The pain, and long confinement, to him worse than pain, he endured with fortitude, and, more than fortitude, with cheerfulness. While he was uncertain of what the event might be, his mind was kept in great anxiety on one point. His eldest son, Lovell, to whose care all that he loved on earth would devolve, was at this time a prisoner in France, where he had been detained for upwards of twelve years. Every possible effort had been made by himself and his friends, to procure his enlargement, or permission to return, even for a short time, to his country and family, but

* Nicholson's Journal for 1817, letter signed Z.

in vain. With great difficulty he had obtained his removal from Verdun to St. Germain-en-Laie, and thence to Paris. All communication, at least all satisfactory communication between him and his family, had been cut off during this long period of captivity. At the time of his father's illness, but before any account of it reached him, an order had issued from the police at Paris to the English prisoners, to quit Paris, remanding them to different places of imprisonment in the country. My brother was of the number, who received this order. This was in March, 1814. The glorious entrance of the allies into Paris changed the fate of individuals, as of nations. My brother obtained his liberty; he was I believe the very first of the English *détenus*, who again touched English ground.

Upon his arrival in London, the first news he heard was of his father's dangerous illness. Without rest he set out for Ireland, and in the shortest possible time reached home. It was late at night when he arrived. My father had been exhausted by a day of great pain, and was just sinking to sleep; but he was completely revived and reanimated by the sight of his son. A few hours, I may say a few minutes, told the history of the thoughts and feelings of years. The ease of mind, the delight my father felt, not only in his son's return, but in finding in that son, as he said, "all that he could wish in his representative, if he were to die, or in his friend, if he were to live," contributed no doubt to the rapidity of his happy recovery. He observed, that here was some trial of the power and efficacy of early education. In the ordinary chances, during twelve years of absence from his home and his native country, a young man's tastes and views in life might be expected to have been changed, and his domestic affections, or at least his domestic habits, altered or destroyed. But, on the contrary, all these appeared, as my father triumphed in remarking, unchanged, and the character was formed and strengthened by the trial of adversity.

One day after his recovery from this illness, when my father was warmly expressing his gratitude to Providence, and his affection to those friends who had attended upon him, he dictated the following:

"I beg it may be remembered, in writing my life, that I now declare, that all the bodily pain I have felt during this illness has been overbalanced by the pleasure of mind I have received from the kindness of my friends, and from the delightful feeling of gratitude."

CHAPTER XIX.

[1815.

DURING the last twenty years of my father's life, his additional experience in education had changed some of his former opinions, and confirmed others. Of the last result of his reflections and observations on this subject he had intended to have given an account in a short "Manual of Education," which he meant to have left as a legacy to his family. With this intention, he wrote the following preface, addressed to his children.

"Since Practical Education was written. Providence has blest me with six children by my present wife, in addition to twelve that I had before. I have attended with care to their education, which has been entirely domestic: I have not determined to follow any preconceived system, but have applied the lessons of my previous experience, to correct what had been useless or faulty in my former endeavors, to educate the elder part of my family, or to confirm what I had found to be successful. I now write in the seventy-second year of my age, and I think it is a duty owing to my children, to let them know the means, which have been taken to cultivate their understandings, to give them a sense of religion, a profound veneration for the unknown cause of their existence, and a sincere and practical submission to those decrees, which are to us in our present state inscrutable.

"I wish to prove to my children, that pains have been taken, to give them moral habits, generous sentiments, kind tempers, and easy manners. How very far my success may have fallen short of my endeavors, it is not for me to determine. But upon one pleasing circumstance in their education I look with confident pleasure; because they, and they alone, are able to judge of it. All my children, except my youngest boy of two years old, are now of an age, when they are able to reflect upon what is past; and they must know, that they have, upon all occasions, been treated with perfect openness and sincerity; that no petty artifices, or what are strangely called pious frauds, have ever been practised, to encourage or deter them; and that they have acquired necessarily habits of truth, not only from precept, but from example. This legacy, then, of a short Manual of Education, I leave to my children, in the pleasing hope, that it will recall to them my memory with sentiments of regard and kindness; and that they will trust, in some degree, in the education of my grandchildren, to the means and precepts, which I have collected, appealing to their own retrospection for the effects, which they have had in the formation of their own characters, * * * * * *."

This fragment is all that remains to us. The loss of the promised legacy, from such a father, is, to his own family, irreparable. To the public, his general sentiments, under the sanction of his own name, are given in two works on education; and I am now to mark only how far these were confirmed, changed, or modified by further reflection, and by the experience of nearly twenty years, which, subsequent to the publication of his first work, he spent in educating another family of children.

More change was made in his opinions on moral, than on intellectual education. As to original genius, and the effect of education in forming taste or directing talent, the last revisal of his opinions was given by himself, in the introduction to the second edition of Professional Education. He was strengthened in his belief, that many of the great differences of intellect, which appear in men, depend more upon the early cultivating the habit of attention, than upon any disparity between the powers of one individual and another. Perhaps he latterly allowed, that there is more difference, than he had formerly admitted, between the natural powers of different persons; but not so great as is generally supposed. His success in his mode of cultivating attention in his new race of pupils was fully as great, and in one instance, I may confidently assert, greater than in any previous attempts. His last pupil, a boy now between nine and ten years old, to whose education his father's mind was devoted, even to the very last days of his existence, and with whom invariably the same methods of cultivation were pursued, is, in the opinion of those who have had opportunities of comparison, the best example, as far as it was permitted to go, of his father's judgment as a preceptor. This boy's memory is good, both recollective and retentive—his judgment clear, his imagination vivid, and his inventive faculty effective; but these different powers of his mind have been kept in such just proportion to each other, that none of them exceed, so as to produce surprise or admiration in the common spectator.

He is so far from being a *show child,* that unless some accidental circumstance called into exercise his power of invention, or brought into view his taste for knowledge, he might pass unnoticed. Less praise, less stimulus of all kinds, were used in his early education, than in that of former pupils; upon the maxim, which applies as well to the mind as to the body, that the least quantity of stimulus, that will preserve it in healthy action, is the best.

The advantage of this mental temperance has in this instance been peculiarly felt. During the course of the last year (1818) at home, he has sustained and employed himself better than could have been expected. His mind has not flagged, as we feared it would, for want of that care and motive, which could not be supplied.

In former days, my father had often been shocked by the long lessons imposed upon young children, and the quantity of time wasted, in a stupid and stupifying manner, in giving common instruction in reading, writing, spelling, &c. He saw, that, by the fatigue induced, children were early disgusted altogether with literature. These observations threw him at one time into a contrary extreme. He did not pay, or allow sufficient attention to be paid, to these ordinary parts of elementary learning. He used to say, that there was no danger, that a child of common sense should not learn to read and spell, and write a legible hand, and obtain a sufficient portion of grammar and history—that these could be secured at any time. Further experience shewed him, that much inconvenience results from the pupils having common things to learn at an advanced and unusual period of education. Though they might excel in higher exercises of the understanding, and possess superior capacity and taste for knowledge, compared with other children, yet there would be a certain sense of inferiority and shame attendant on the consciousness of ignorance, even of things perhaps not in themselves valuable. This shame, besides being a disagreeable feeling, is to be dreaded at the time when young people are entering into the world. It embarrasses them just at the moment, when the mind ought to be free and vigorous for the competitions of youth. With pupils who have had a private education, this is particularly dangerous. When they first measure themselves with others, if they be made to perceive any deficiency obvious to ridicule, it tends to raise a fear, that all is wrong in their education; and they think, that, in the midst of the hurry and pressure of immediate business, there can be no time or possibility for remedy. Longer experience in the world may perhaps calm these fears, and restore confidence in themselves, and in the more valuable part of their previous education; but why run the hazard? My father with his last pupils found the advantage of having the common elementary knowledge taught early and securely. He became sensible, that more of what may be called *drudgery of mind*, than he had formerly thought advantageous, is not only useful, but necessary for children, to train them to that degree of application, to which the quickest talents must submit, before they can succeed in any profession, or before they can advance in any path of business, science, or literature; crowded as every path now is with competitors, even genius is doomed to *labor* before it can succeed.

I hope to be clearly understood, that my father rectified errors, but did not change or abandon any of the principles, which he recommends in cultivating the mind. Though he was convinced, that greater and more continued habits of application were necessary, than he had formerly imagined; yet in training to the power of application, he pursued uniformly the same methods.

With respect to public and private education, he was confirmed in his decided preference of private education for girls; but for boys he latterly never recommended private tuition, except when there is a concurrence of favorable circumstances, which I fear cannot often happen. This I find strongly expressed in his own words, in a letter written in 1812, to the Editors of Rees's Cyclopedia. After speaking of some favorable circumstances*,

---

* "You speak of my daughter, as if she had been practically employed in education. This is not the case. She is my pupil, my literary partner, and my friend. But I have not exacted from her any laborious interference in the education of my numerous family. Her constant attention to all of them is undoubtedly of the utmost service; and her strong attachment to my wife gives a unanimity to all our pursuits,

which had enabled him to pursue with his large family a course of private instruction, he adds,

"I most earnestly deprecate the conclusion that has been drawn from our books, that we recommend in general private education for *boys*. We know, that in general private education is impracticable, and that it requires an uncommon coincidence of circumstances, to make it in any case advisable."

In Practical Education the value of a good memory is perhaps underrated. My father possessed this faculty in a superior degree, both retentive and recollective; but this power, in comparison with those of invention and reasoning, he had always, with provoking raillery, taken pleasure in depreciating. As he advanced in life, he felt the advantages of memory in his own case, and became more sensible of its being essential to the exercise of the other faculties of the mind. When his sight began to fail, and to render reference to books less easy to him, he found, even in the exercise of his reasoning and inventive power,

---

that has a most salutary effect. I have also in my house two sisters of my last wife, whose kindness and attention have always tended to the same objects in the education of my family. With these advantages, it is easy for me to accomplish, what in other circumstances would be morally and indeed physically impossible."—*Extract of a Letter to Dr. Rees.*

the blessing of memory, and the advantage of having his own well-stored, ready to supply him with materials for thought, and for independent occupation. Besides his having formerly undervalued the faculty, he had despised accuracy in registering knowledge of facts, names, and dates. But, subsequently, he observed the effect and utility of a contrary practice. He perceived, that a recollection of names and dates may, in some happy exceptions from the common laws of mind, be found in combination with the higher intellectual powers, without injury or incumbrance to the understanding; and that the taste for better knowledge may have actually survived the labor necessary for the acquiring these habits of careful registry.

In one example, daily recurring to him in some agreeable or useful instance, he felt, both in the pleasure of conversation, and in the conduct of business, the advantage of accuracy of memory. With his characteristic candor, he delighted to acknowledge this, and he acted immediately in consequence of his conviction. He took successful pains, to improve the accuracy of his own memory, and he did the same for his pupils. Still, however, adhering to his former principles, he held it desirable only so far as it can minister usefully or ornamentally to the higher powers of the mind.

I am not aware of any other material alterations in his practice or principles, with respect to *intellectual* education. With regard to *moral* education it is of consequence, to mark strongly the result of his enlarged and longest experience.

In his earliest attempts he had endeavoured to reduce to practice Rousseau's theories. Finding the bad effects, which resulted from following this system, from trusting too much to nature, liberty, free will, and the pupil's *experiments* in morality, my father for some time afterward inclined to the extreme of caution; and became more apprehensive than was necessary of the effect of trifles, or of small, accidental temptations. This perhaps appears in the first edition of Practical Education, in the chapters on "Acquaintance" and on "Servants."

Further experience convinced him, that it is impossible in the world in which we live, to exclude from the sight, hearing, and *imagination* of children, every thing that is wrong; that the seclusion necessary for the attempt would be not only difficult, but dangerous, because it would leave the judgment and resolution uninformed and unexercised on many points of conduct and manners. He found, that the best chance of avoiding danger from bad example is to give, as early as possible, means of comparison, and habits of resolution, to let young people see different characters, conduct, and manners, and hear, in the course of common conversation in family society, the opinions that are given, without reference to themselves. This exercises them early in that sort of resolution, which is found necessary in real life, to enable people to refrain from imitating what is affected in manner, or wrong in conduct.

In consequence of this conviction, he became less apprehensive of the effect of casual example, and of small, external circumstances. He was convinced, that his children were not hurt, by seeing a greater variety of people. On the contrary, he found its advantage in a certain robustness of mind, and security of mental health, which cannot be obtained without a good deal of free exercise, and some hazardous exposure.

As to servants, it should be remarked, that in his childhood and youth the race of servants in Ireland was peculiarly bad. Facts, which I have heard from him and from persons of his time, convince me, that some of those servants were the most unprincipled and horrible companions imaginable for children. Some circumstances which he saw in his childhood, some which afterwards in his youth came to his knowledge during his acquaintance with Sir Francis Delaval, deeply impressed his mind with a sense of the danger to young people of associating with such servants as he then knew. Hence his cautions were vehement to parents, and his regulations on this point even more strict than necessary, or than in many families could be feasible. His conviction of the general danger remained strengthened by all his observation, that intercourse with bad servants is one of the secret reasons of the frequent failure of otherwise careful and apparently good education among the higher classes. He believed, that ladies' maids, young gentlemen's gentlemen, footmen, grooms, and coachmen, have often more influence, than the preceptor or the governess; and that they are in fact, remotely or immediately, the cause of much of the extravagance, vice, and follies of those, under whom they seem to serve. But though this conviction continued, and though my father spoke with unabated force upon the subject whenever it was questioned, yet he in his own family, during the last twenty years, did not find occasion for that extreme vigilance, and those *ultra* precautions, on which he formerly insisted. The habits of the family were so far established, that no extraordinary care was necessary. Without any prohibitions, the boys never thought of talking to the men-servants, or of making them their play-fellows or companions. The children were so completely part of our society, and so much entertained and interested in what was going on among us, so much at ease with us, that there was no temptation for them to go to the servants in search of amusement, or to escape from constraint. In reality, these are the best and only perfect securities.

I have stated the facts, and I have only to add, that, as far as I can judge, no evil has ensued. On the contrary, the freedom from all appearance of unnecessary vigilance, and of small regulations or restraints, must increase the confidence of the children, and prevent them from feeling, what the pupils of private education should never feel—*that they have been managed in any peculiarly strict manner*.

There is another point, on which I may with confidence give the result of my father's latest experience. Formerly, from having observed how apt children are to dispute and quarrel, when they are left much together, and from the fear of the strong becoming tyrants, and the weak, slaves, it had been thought prudent, to separate them a good deal. It was believed, that they would consequently grow fonder of each other's company, and that they would enjoy it more, as they grew more reasonable, from not having the recollection of any thing disagreeable in each other's tempers. But my father became thoroughly convinced, that the separation of children in a family may lead to evils, greater than any partial good that can result from it. The attempt may induce artifice and disobedience on the part of the children; the separation can scarcely be effected; and if it were effected, would tend to make the children miserable. He saw, that their little quarrels, and the crossings of their tempers and fancies, are nothing in comparison with the inestimable blessings of that fondness, that family affection, which grows up among children, who have with each other an early and constant community of pleasures and pains. Separation as a punishment, as a just consequence of children's quarrelling, and as the best means of preventing their disputes, he always found useful. But, except in extreme cases, he had rarely recourse to it, and such seldom occurred. In short, upon this and other points, which I need not enumerate, the greatest change, which twenty years further experience made in his practice and opinions in education, was, to lessen rather than to increase regulations and restrictions. He saw, that, where there is liberty of action, one thing balances another; that nice calculations lead to false results in practice, because we cannot command all the necessary circumstances of the data. In the course of a long life, he saw many children grow up to men and women, and saw the result of much careless and careful education. He found, that many turned out well, notwithstanding certain omissions or deficiencies in their course of tuition.

Instead of becoming, as he grew older, more pertinacious in pursuing his own way, he was more ready to allow others to follow theirs—secure that they might, by different paths, attain to the same objects. As experience extended his views, he made more allowance for the short-sightedness of others, and became more and more tolerant; but he never even inclined to be in the least sceptical with respect to the power of education, or the certainty of its action. Quite the contrary; his belief in its power increased and strengthened to the end of life.

As his anxiety and exactness about the less points decreased, he grew more eager and attentive to strengthen the great moral principles of action.

For many years of his life he had, I think, been under one important mistake, in his expectations relative to the conduct of his fellow-creatures, and of the effects of cultivating the human understanding. He had believed, that, if rational creatures could be made clearly to see and understand, that virtue will render them happy, and vice will render them miserable, either in this world or in the next, they would afterwards, in consequence of this conviction, follow virtue, and avoid vice. He conceived, that in the multitude of instances, in which this seems not to have been the case, he could account for it by various causes. He thought there might have been some confusion in the mind, from the use of the abstract terms, *virtue* and *vice;* or from the person's not having verified the moral demonstration accurately; or from assent being given in words merely, without the conviction having reached the understanding; or from that want of experience, which prevents the possibility of an individual comparing all those sorts and degrees of pleasure and pain, which constitute happiness or misery.

Hence, both as to national and domestic

education, he formerly dwelt principally upon the cultivation of the understanding, meaning chiefly the reasoning faculty as applied to the conduct. But to see the best, and to follow it, are not, alas! necessary consequences of each other. Resolution is often wanting, where conviction is perfect.—Resolution is most necessary to all our active, and habit most essential to all our passive virtues. Probably nine times out of ten, the instances of imprudent or vicious conduct arise, not from want of knowledge of good and evil; or from want of conviction, that the one leads to happiness, and the other to misery; but from actual deficiency in the strength of resolution; deficiency arising from want of early training in the habit of self-control. Another circumstance is to be remarked, that, under the influence of any passion, the perception of pain and pleasure alters as much as the perceptions of a person in a fever vary from those of the same man in sound health. The whole scale of individual happiness, as well as of general good and evil, virtue and vice, is often disturbed at the very rising of the passion, and totally overthrown in the hurricane of the soul. Then, in the most perilous and critical moments, the conviction of the understanding is, if not reversed, suspended. Those who have lived long in this world, and who have seen examples of these truths, feel that these are not mere words. Such are the points, which I think my father would have urged with the utmost force, if he had written his last sentiments on Education.

The silence, which has been observed in Practical Education on the subject of religion, has been misunderstood by some, and misrepresented by others.

To misrepresentation apparently wilful, and made in the acrimonious language of party-spirit and intolerance, my father never deigned to reply. But to those, who with upright and benevolent intentions, from a sense of public duty, and in a spirit of christian charity, made remonstrances on this subject, he thought it due to give all the explanation in his power. Professor Pictet, of Geneva, was the first, to whose able and kind animadversions in the Bibliothéque Britannique he replied. The correspondence, which passed between them, is prefixed to a French translation of "Practical Education," published at Paris and Geneva in 1800.

In the preface to the second English edition, published in 1801, my father, by whom that preface was written, adverts to M. Pictet's strictures, and concludes with these words—

"The authors continue to preserve the silence upon this "subject, which they before thought prudent; but they *disavow,* "*in explicit terms, the design of laying down a system of* "*education founded upon morality, exclusive of religion.*"

After this declaration, my father never felt himself called upon to say more upon the subject, till, in 1812, he accidentally met with an admirable article on "*Intellectual Education*" in Rees's Cyclopedia. As we read on, we found in it many expressions so flattering to ourselves, that I should not here venture to mention it, but for the importance of the subject, on which it is necessary to cite the facts. After speaking of "Practical Education," and of some of our elementary works, the writer of this article, in terms the most delicate, but at the same time the most distinct, and the most honorable to his own piety, deplores "their striking and ever to be lamented deficiency in every thing like religious principle."

My father instantly wrote to the editor of the work the following letter:

"SIR, *Edgeworth-Town, Aug.* 18, 1812.

"Having this day seen the article, '*Intellectual Education*' in your Cyclopedia, I cannot delay returning my acknowledgments for the very candid and liberal manner, in which 'Practical Education' has been discussed in your dictionary.

"The large mind, which comprehends and directs the magnificent national work, in which you are engaged, is every where visible; and we consider, that what would have been only of temporary duration in our book, will now be permanently preserved in yours. I must, however, regret, that an error pervades the whole, which has been adopted by most of our critics, and which we most earnestly deprecate—the imputation of disregarding religion in Education.

"In the French translation of 'Practical Education,' this subject is discussed in the preface, and I beg from your justice, that some occasion may be taken of entering our protest against this charge. We hope, that in 'Professional Education,' under the head 'Clerical Education,' we have evinced a proper sense of the Clerical character, and an enlarged view of religious sanction. This chapter was written for the Clergy of the Establishment, to which we belong; but our views in 'Practical Education' were not confined to any sect or nation.

"Our private tenets are of little consequence to the public, but we *are convinced, that religious obligation is indispensably necessary in the education of all descriptions of people, in every part of the world.*

"We dread fanaticism and intolerance, whilst we wish to hold religion in a higher point of view, than as a subject of seclusive possession, or of outward exhibition. To introduce the awful ideas of God's superintendence upon puerile occasions, we decline. At the same time, we have not presumed to blame others for acting upon a different persuasion.

"I have the honor to be a member of the Board of Education in Ireland. My opinions on the subject of national education appear in our reports. By these I hope I shall obtain the justice due to me on the subject, and that it will appear, that *I consider religion, in the large sense of the word, to be the only certain bond of society.*

"You have turned back our thoughts to this most important subject (education), upon which, *next to a universal reverence for religion,* we believe the happiness of mankind to depend. We shall in consequence write a short tract, in which we shall endeavor to evince, that the greatest favor, which an author can receive, is a candid examination of what he has written, and a fair investigation of his errors."

I may be permitted to add, that many distinguished members, and some of the most respected dignitaries of the Established Church, honored my father by their esteem and private friendship. This could not have been, had they believed him to be either an open or concealed enemy to christianity; or had they conceived it to be his design, to lay down a system of education founded upon morality exclusive of religion. As a member of the Board of Education, and as a member of Parliament, he had public opportunities of satisfactorily recording his opinions; and as a magistrate for a long course of years, he supported these opinions by his actions, by his example, and authority.

I have often been witness of the care, with which he explained the nature and enforced the observance of that great bond of civil society, which rests upon religion. The solemnity of the manner, in which he administered an oath, can never leave my memory; and I have seen the salutary effect this produced on the minds of those of the lower Irish, who are supposed to be least susceptible of such impressions. But it was not on the terrors of religion he chiefly dwelt—no man could be more sensible than he was of the consolatory, fortifying influence of the Christian Religion, in sustaining the mind in adversity, poverty, and age. No man knew better its power to carry hope and peace in the hour of death to the penitent criminal. When from party bigotry it has happened, that a priest has been denied admittance to the condemned criminal, my father has gone to the county gaol, to soothe the sufferer's mind, and to receive that confession, on which, to the poor Catholic's belief, his salvation depended. Whatever their peculiar tenets might be, none of his fellow-creatures, in any rank of life, or in any connexion or relationship to him, of servant, tenant, dependent, friend, were ever by him disturbed in their faith; nor did he ever weaken in any heart, in which it ever existed, that which he considered as the greatest blessing that a human creature can enjoy—firm religious faith and hope. No man could be more tolerant than he was, in judging of the religious opinions of all classes and sects; provided he had reason to think them sincere, he shewed them the utmost respect; and by such conduct he did more good to the cause of religion, of virtue, and civil peace in Ireland, than could have been effected by the most rigid disciplinarian, or by the most furious zealot.

CHAPTER XX.

" But grant the virtues of a temperate prime
Bless with an age exempt from scorn or crime,
An age that melts in unperceiv'd decay,
And glides in modest innocence away;
Whose peaceful day benevolence endears,
Whose night congratulating conscience cheers,
The general fav'rite, as the general friend:
Such age there is, and who could wish its end?
Yet, ev'n on this her load misfortune flings,
To press the weary minute's flagging wings;
New sorrow rises as the day returns,
A sister sickens, or a daughter mourns;
Now kindred merit fills the sable bier,
Now lacerated friendship claims a tear;
Year chases year, decay pursues decay,
Still drops some joy from with'ring life away;
New forms arise, and diff'rent views engage,
Superfluous lags the veteran on the stage
Till pitying nature signs the last release,
And bids afflicted worth depart in peace."

THESE are beautiful lines, but this picture of age, drawn by the master-hand of a great moralist, is darkened too deeply by his constitutional melancholy. If the virtues of a temperate prime include energy of character, and habits of useful activity, age may hope to be blessed with more than the negative happiness of being exempt from scorn or crime, something more even than melting in unperceived decay.

In the course of his life, my father had known the weight of misfortune; he had felt what it is to lose the object of the fondest affection; he had submitted to the loss of children, and to the death of friends; year after year, sorrow after sorrow had pressed upon him; but, instead of suffering his mind to fix in ungrateful despair upon the joys, which "drop from withering life away," he had, with true gratitude to Providence, enjoyed the blessings which remained, and had resolutely cultivated the promise of another spring. New forms arose, and different views engaged; but these forms were dear, and attached to him by grateful affection, and in all these different views he sympathised and assisted. Never could he know the dreadful feeling of the veteran, who lags superfluous on the stage; for besides his public use and influence in a wide circle of connexions, he was conscious, that his large family were continually guided by his experience, excited by his energy, and made happy by kindness, such as a father only can feel or bestow.

When he was about sixty, I recollect his being struck with an expression used by a friend much farther advanced in years than himself, the Abbé Morellet.

"I hear people complaining of growing "old, said the venerable Abbé, but for my "part, I enjoy the privileges and comforts, "in short, the convenience of old age, (*les* "*commodités de la vieillesse*)."

This amiable, respectable, and respected old man, in some playful lines he wrote on his own birthday, declares, that, if the gods were to permit him to return again on earth, in whatever form he might chuse, he should make perhaps the whimsical choice of returning to this world as an old man. Without going quite so far as this, my father was, as he advanced in years, as cheerful, and more serenely happy, than at any previous period. Reflecting upon and comparing his felicity at different ages, he often assured me, that he thought the last thirty years of his life had been the happiest. It is remarkable, that though men, as they advance in years, seem so much to regret the past, yet there are few, who, if the offer were made to them, would like to live their lives over again. Of the numbers to whom my father proposed this question, I have never known any person answer in the affirmative. Him alone

I have heard repeatedly declare, that, provided he were allowed to correct some faults, he would willingly accept the terms. He could not more strongly mark his conviction, that the sum of good he had enjoyed infinitely surpassed the pain or evil. In his latter years, his children often heard him repeat, that he had been happy in this world beyond his utmost expectations, and that he really thought he had been allowed more than his share of blessings. His heart, instead of contracting or sinking, seemed in age, to expand and overflow with kind feelings towards his fellow-creatures, and with gratitude to his creator.

I particularly recollect his expressing these sentiments to us one evening, when he was walking out for the first time, after recovering from that painful illness, which it has been mentioned he had in his seventy-first year. It was a fine evening in autumn. After repeating those lines of Milton's, which recur probably in similar circumstances to every feeling and cultivated mind:

———————— " Fragrant the fertile earth,
After soft show'rs; and sweet the coming on
Of grateful evening mild:"

he said, that, as he had grown older, he had become more sensible than he ever was in his youth to the beauties of nature, to the pleasure of observing the seasons and their change, to the influence of cheerful sun, and refreshing breeze, to the song of birds, and to all those simple pleasures, which are commonly supposed to be peculiarly felt by the young, and to pass away with the first impressions of novelty, or with the first enthusiasm of poetry and romance.

The being capable of returning to these simple pleasures, and enjoying them in age, is, as he remarked, one of the rewards of those, who preserve, with a peaceful conscience, minds untainted by avarice or ambition, and undisturbed by any malignant passions.

At this moment, when he was enjoying renovated health, he observed, that this pleasure never could have been felt by a being, who had not known the taste of pain. Adverting to Burke's use of the word *delight,* which in the Essay on the Sublime and Beautiful is defined to be the feeling, that arises from the cessation of pain, he pointed out to us, that, though it is the lot of our nature to *suffer,* yet even thence beneficent Providence has ordained, that a new pleasure shall arise.

One evening after having repeatedly said, "how I enjoy my existence!" he observed, that Dr. Darwin was right in asserting, that a sense of pleasure is annexed to the mere feeling of existence; and that this, independently of other causes, creates an attachment to life, increasing with habit, instead of diminishing and fading, like other enjoyments, with the charm of novelty.

From the same habit of belief, which leads us to expect, that the sun which rose yesterday will rise to morrow, old men continue to believe, that they shall live to morrow, as they have lived to day. And though increasing infirmities, or the frequency of the deaths of those of the same age, warn the old, that they cannot last beyond a certain term; yet the mere irrational habit prevails so far, as to counteract much of that apprehension, which might otherwise embitter the latter years of life.

These things my father pointed out to us, as some of the beautiful provisions, which have been made in our nature for the tranquillity of age. Whatever he said on such subjects came with peculiar force from one, who was so free from all affectation, that his words carried with them perfect conviction of the truth of his feelings.

The same moralist, who described the de

cline of life in such melancholy strains, has said, that no man adds much to his stock of knowledge, or in short improves much, after forty. Of this assertion my father always doubted the truth, and he opposed the principle, as injurious to the cause of knowledge and virtue, and tending to lessen the energy and happiness of a large portion of human existence. From forty to threescore years and ten, the ordinary limits of the age of man, comprises above a third part of his life, during which his intellect would be by this principle condemned to remain stationary, or to become retrograde. Every instance to the contrary is encouraging and useful to humanity; and such my father loved to recollect and record; his own example may now be added to the number.

As the preceding account of his life shews, it was during the last thirty years, from forty to seventy, that he exerted himself most constantly and effectually in useful public undertakings, contributed most to literature and science, and most improved in the art of education; the subject to which he had from early youth devoted the greatest share of his attention. Above all other excellencies he valued for himself, and for his friends, this power of improving. He used to say, that with it they might in time be any thing, and without it in time they would be nothing.

One day, in searching with my father among some old papers, I found some sheets of a commonplace book, which he had kept when he was a very young man. He looked at the pages with an expression of extreme pleasure in his countenance, for, as he said, they recalled to him most agreeable associations with the time*, when he first strongly felt the delight of voluntary study, when he could say at the end of each day, that he had added something new to his stock of knowledge, or had in some respect improved himself. The same habit of mind continued through life. He was proud of shewing, that he could in advanced age acquire new knowledge, or correct any habit or fault, trifling or serious. He often assured his children, and, more than asserted, demonstrated by his daily practice, that it was his ambition to improve, even to the latest hour of life. To this desire early formed he ascribed a great portion of his peculiar happiness in age, a season of life for which he had long prepared before he began to grow old. Without affecting to be

---

* See his own Narrative, vol. i. p. 91.

young, he exerted himself to prevent any of his faculties from sinking into the indolent, torpid state, which precedes and portends their decay. The impression which had been made on his mind, as I have mentioned, by some passages on this subject in the life of Reid, became highly useful to him in practice, making him watchful over his own mind, and sedulous even of what to the ignorant and careless might appear trifling habits. He assisted his memory so much by reasoning and arrangement, strengthened it so much by exertion, that it became in age more alert and vigorous even than it had been in youth.

He particularly enjoyed the society of those, who had the habit of recurring to classical literature, and to the studies of his youth. This, I observed, encreased in a high degree the pleasure he took in the company of one of his earliest friends and school-fellows*, whose conversation is rich and eloquent in classical allusion.

Sometimes he amused his mind with arithmetical problems, such as required at once memory and ingenuity. His inventive faculty he exercised with equal success.

Independently of the positive value or use

* The present bishop of Meath.

of the knowledge obtained, there was continual pleasure in the effort, and in the motive. Next to the perception of progressive improvement, the sense of the endeavor to improve is satisfactory. It certainly assists what the poet so well calls "congratulating conscience."

In his latter years, my father reflected much on the course of his own mind, and considered carefully what circumstances had tended to form his character, observing how far his own conduct, and the habits of his understanding, had influenced his happiness.

As every human being, if he live, must grow old, it cannot be uninteresting to know the opinions of one, who had seen much of the world, and reflected still more on his own life and feelings. There *can* be nothing new in any moral truths, nor in any mathematical demonstration: but he is assuredly serviceable to the cause of virtue, who adduces even one additional example in its favor, and who offers one evidence more in the affecting form of testimony, left by the dead to the living.

To what he has left in writing I may now add some of the observations on his own mind, which I have heard from him in conversation, and some which long intimacy and perfect confidence have enabled me to make.

It appears from his Memoirs, from his subsequent letters, and from this narrative, that his scheme of happiness, and manner of life, continued the same during nearly half a century, from the age of five and twenty to seventy. How few can look back through such a length of time, and feel, that, though the boy and man an individual make, their past and present selves make one and the same consistent person! Such consistency is the more remarkable in one, whose extraordinary vivacity might have led us to expect changeability in conduct, proportioned to versatility of imagination. Much of the steadiness of his course through life may be attributed to his having in youth turned his attention to the observation of his own feelings, and of the circumstances, external and internal, which, independently of fashion or prejudice, formed the real happiness or misery of all those, with whom he became acquainted. He also continually rejudged and discussed his observations, with his moral, and, in the best sense of the word, *philosophical* friend; his uncommon candor, in acknowledging those changes of his own opinions and feelings, which resulted from the conviction of his understanding, enabled him really to profit by *philosophy* and *experience*, which are often mere words sounding on the ear, but not operating on the conduct. Even those who determine wisely what is essential to their happiness, and who have sufficient resolution to abide by their plan of life, sometimes fail in adapting the means to the end. These, finding themselves unhappy, blame their connexions, friends, accidental circumstances, or peculiar ill-fortune, for that which arose from some defect in the management of the minor parts of their plan; or more commonly from some error in the management of their own minds. For instance, it has not unfrequently happened, that men who love domestic life, and who are extremely attached to their wives and families, have, nevertheless, rendered those connected with them wretched by want of temper, by disagreeable manners, or by that neglect of personal appearance, which tend ultimately to repel and disgust affection.

After having made domestic life his choice, it was my father's constant care, to attend to all that could make domestic life happy; for this purpose, he applied to the management of his mind all that he observed in life, or that he met with in books.

Among the ponderous tomes of casuistry,

and of false or wordy philosophy, which lie on the shelf unopened, and useless for the practical purposes of improvement, there is one little volume, which my father valued highly, not for its literary composition, for in point of style and arrangement it is very defective, but for its use in the management of the mind. Locke's short "Essay on the Conduct of the Understanding." He profited by many of its suggestions, in guarding against the faults and erroneous habits, which it points out. But this book is confined chiefly to intellectual improvement. My father pursued a similar plan in marking the slight causes of moral error.

He observed, that the happiness, that people derive from the cultivation of their understandings, is not in proportion to the talents and capacities of the individual, but is compounded of the united measure of these, and of the use made of them by the possessor; this must include good or ill temper, and other moral dispositions. Some, with transcendant talents, waste these in futile projects; others make them a source of misery, by indulging that overweening anxiety for fame, which ends in disappointment, and excites too often the powerful passions of envy and jealousy; others, too humble or too weak, fret away their spirits and their life in deploring, that they were not born with more abilities. But though so many lament the want of talents, few actually derive as much happiness as they might from the share of understanding, which they possess. My father never wasted his time in deploring the want of that, which he could by exertion acquire. Nor did he suffer fame in any pursuit to be his first object.

Far beyond the pleasures of celebrity, or praise in any form, he classed self-approbation and benevolence; these he thought the most secure sources of satisfaction in this world. This truth he *suspected*, as he said, in the beginning of life; and of this he became certain, from experience, in the end.

When he was a youth of nineteen, an old gentleman, who saw him passing by his window, said of him, judging by the liveliness of his manner and appearance, "There goes a young fellow, who will in a few years dissipate all the fortune his prudent father has been nursing for him his whole life."

The prophecy was by a kind neighbour repeated to him, and, as I have heard him say, it made such an impression, as tended considerably to prevent its own accomplishment.

He acquired the habit of calculating and forming estimates most accurately. He not only estimated what every object of fancy and taste would cost, but he accustomed himself to consider, what the actual enjoyment of the indulgence would be. Carefully analysing his own feelings, he observed how much, or rather how little, our own real tastes are gratified by the purchase or pursuit of many things, to which we are induced by the example of others, or by the desire to possess what others admire. To counteract this sort of sympathetic effect on the imagination, he upon all occasions carefully separated the idea of the pleasure of possession from that of contemplating any object of taste.

In his narrative of his youth he has said, that he never indulged in any expensive taste for pictures, books, or ornamental building. I may go further: I may safely assert, that he never threw away in the course of his life twenty guineas for himself on any ingenious or luxurious knicknack. This may give some standard, by which, according to the common mode of estimating, we may measure the strict order, in which his prudence kept his fancy.

Persons of imagination and invention are

apt to enter into speculations and projects. He never speculated; and in pursuing his favorite taste in mechanics, even when his enthusiasm was raised by the hope of executing some invention or work useful to his country, and honorable to himself, no project ever led him into what his family, or friends, or any impartial judge could think imprudent expense. This constant attention, to keep his imagination within prudential bounds, certainly did not, as he advanced in years, lead to selfishness. In every habit of his mind, every emotion of his heart, he was as generous as in youth—more truly generous, because he acted from more enlarged and enlightened views.

It is the complaint of poets, and theme of the weeping class of philosophers, that Hope ever tells a flattering tale, which is continually followed by disappointment, and that all human pleasures are greater in the expectation than in the reality. Aware of the danger, he listened with caution to Hope's flattering tales, comparing them constantly with the records of experience, and expecting no romantic or miraculous exceptions in his own favor. He never amused himself in what is called *castle-building;* on the contrary, he discouraged the painting, or dwelling on the contemplation, of fancy pieces of felicity.

Even when almost certain of any future pleasure, he tried to occupy his attention, as far as possible, with other objects, on purpose that he might not waste or weaken his feelings by anticipation. In consequence of these precautions, instead of being disappointed and discontented with himself, with his friends, and the world, he found through life, as he repeatedly assured us, every real good greater in the possession, than in the expectation.

In the management of his temper, as much as in the government of his imagination, he made the most advantageous use of his understanding. His temper was naturally warm, and, as he has said, violent. Upon trifling occasions he might speak hastily; but when he felt serious displeasure, he became silent, still, and composed; all the powers of his body and soul seemed to be concentred in bearing the pain, and restraining the passion. He had imposed upon himself this habit of suspension from decision or action, lest he should act or decide rashly. He found this immediate and absolute restraint and silence the most efficacious means of conquering the passion, giving time for the return of reason, and gaining the victory by delay. It has been quaintly said by Sir Francis Bacon, that "an angry man, who suppresses his passions, thinks worse than he speaks; and an angry man, who will chide, speaks worse than he thinks." On this principle, many dread the angry man who is silent, and prefer the temper which lets out at once all its violence, rather than the silence from which they apprehend deliberate and lasting resentment. But it is of essential importance, to distinguish between that prudent and honorable silence, which a wise and good person imposes on himself, lest he should do any wrong or injustice, from that base silence, which is held with the malicious intent, and with all the mental reservation of hatred. Unless we make this distinction, we shall, as my father observed, prevent the young and generous, the warm tempered and warm hearted, from using the first and safest precaution against the passion of anger, and we shall encourage the yielding to all its caprices and fury; for who would not rather be thought passionate, or even furious, than be suspected of dissimulation or malice?

In truth no one could suspect my father of any kind of mental reservation. All who lived with him were secure, that he never was displeased, without letting them know it. If they were in any doubt as to the cause, they had only to ask, and were sure of hearing, not the evasive replies of "Nothing of any consequence,"—or "Why should you think I was displeased"—but the truth, the plain truth, and the whole truth. So that every dependant, or inferior, was certain of being allowed full opportunity of vindication and justice; and every friend, time for explanation, with a heart open in their favor. He used to say, that all he asked of his friends was to suppose, that he meant kindly, till it was proved, that he meant unkindly; and he did the same by them.

This sincerity he thought not only necessary to the general confidence, that ought to subsist in a family, but beneficial to the temper. Those who have the courage to bear the shame of telling the petty causes of their displeasure, or who, in more important cases, can own themselves in the wrong, will, he thought, have the resolution to correct any fault of temper; and will, by such candor, be secure of holding the esteem and regard of every friend worth preserving.

But let it be clearly understood, that the sort of explanation, which he recommended and practised, was not a recriminatory inquiry, held merely to find out who had been most to blame on any particular occasion, but an examination into the grounds and general principle of the cause of offence: it was not for the purpose of useless, and therefore, cruel reproach concerning the past, but of wise and kind precaution for the future. The truth and the reason of the case being made to appear to the judgment when cool, there was no danger, that these explanations should be resorted to so frequently as they sometimes are, where the object is only to set pride against pride, and passion against passion.

If in any moment of sudden anger—and to such the best and wisest, notwithstanding all their precautions, are subject—if in any sudden moment of anger, my father was unjust, even in the slightest degree, he never rested, till he had made acknowledgment and reparation, even for a hasty word, especially if it were to an inferior, a dependant, or a child.

But before he had reached middle age, he had acquired the habit of growing cool in discussion or dispute, when he saw his opponents grow angry. He seemed to be put on his guard by the appearance of that passion, of which he well knew the symptoms, and he allowed for the infirmity; indulgent to others, while severe to himself.

And what was the result of this?—the best result, and the best reward which he could have in this world, domestic felicity. He who lived so well with so many different dispositions, and obtained over so many such strong influence and devoted attachment, must have had a good temper; but there was, in fact, an extraordinary *care*, as well as natural happiness. I have said, that he never lost a friend, except by death. This part of his happiness must be in a great measure attributed to his well governed temper; for it is well known by all who have any experience, that the best heart and the most generous disposition, the highest talents and deepest capacity, are, none of them, alone or combined, sufficient without temper to preserve friends.

To the fact that he never lost a friend, I should add, that in some cases where crossing interests, previous family quarrels, and long lawsuits, had made enemies of neighbours, his command of temper in business, his candor and generosity, whether conquered or conquering, completely won his adversaries. Even in a solitary instance, where temperate and generous conduct during long years of lawsuits seemed to have been of no avail, it was ultimately successful in preventing the spirit of enmity from descending to another generation. The sons were won by the conduct, which the father could not appreciate. The handsome manner, in which they have shewn their sense of respect and regard, where they felt that it was due, does honor to themselves, and in the decline of life gave benevolent sutisfaction to him, who, far beyond any selfish triumphs, delighted in finding good in human nature, and in promoting the peace and happiness of all within his sphere of influence.

There are many who in youth, and perhaps till middle age, persevere in attempts to improve their tempers, but afterwards indulge themselves, if indulgence it can be called, in the idea that they are growing too old to mend; or they quiet their consciences with the assertion, that *such is their natural humor*, and their friends must take it for better for worse. Hence the temper usually relapses in age into all the faults, from which it had been reclaimed during the season, while the tender or the interested passions supplied motive for vigilance. But my father never considered temper as a positive quality, that was to remain unaltered after some given period of life. He was sixty when he wrote his memoirs, and during his succeeding years, I can answer for it, that he never relaxed in his attention to the regulation of his temper. The consequence was, that it meliorated, instead of degenerating in age: it continued to be warm, but it was always open to conviction; gentle the moment after anger, placable, generous; generous, especially in forgiveness; affectionate to enthusiasm, and tender beyond what any one, judging either from his serious tone of philosophy, or from his gaiety in conversation, would have suspected.

But I must check myself.

One day, when I was most anxiously intent upon writing these memoirs, I opened on the following sentence in the life of a great man of antiquity.

"*His danger arose from the praises bestowed upon him by those worst of enemies, the dealers in panegyric.*"

These words seemed like a warning to me, and so I felt them; not superstitiously, but rationally. They have operated strongly, to restrain the dangerous impulses of affection and gratitude. I have felt myself incapable of general reflections or deductions, such as adorn and illustrate biography, when written by a master-hand, and by a mind at ease; yet impartial, as far as possible, I hope that I have been; and I trust that my scrupulous adherence to truth, and my intimate knowledge of the character I draw, may perhaps compensate for other deficiencies.

To the public the great difficulty is, to obtain the full and exact truth respecting the education and private feelings of individuals, and to learn the result of their experience, and their own reflections on the events of their lives. Unless these be accurately ascertained, biography, even when written in the most elegant style, and with the most elaborate display of metaphysical or moral reflectio, is all delusion or uncertainty, and from its perusal no consequence useful to mankind can result: but, if the facts be fairly and fully stated, they present valuable materials, from which all may draw their own conclusions, and on which minds of various and superior powers may afterwards work.

From the true history of individuals in different classes of society various results may be expected. The effect of the lives of heroes, warriors, and statesmen, is to inspire the love of military glory, and "to make ambition virtue." Compared with the great mass of mankind, there can however be only a small number of heroes; nor does it appear certain, that, even when most successful, they are, computing the whole of life, much happier than other men.

The history of their lives, however, exalts the human mind; induces others to follow the same course; and thus raises, from time to time extraordinary beings, necessary to the state in great emergencies, sacrifices to the good or to the pride of their country.

The biography of private individuals has a more humble, but not less useful object—to improve mankind in the social and domestic virtues, by shewing how much these tend to human felicity.

In this point of view, these memoirs will it is hoped be useful to the public, and pursuing this object, the biographer must, however painful, speak of the last years of this well spent, happy life.

CHAPTER XXI.

THOUGH my father had to all appearance recovered from his dangerous illness in 1814, yet he frequently reminded us, that, at his advanced time of life, he could not for any length of time expect to remain free from sickness, or from the infirmities of age. He did not for his own sake desire length of life; but it was his prayer, that his mind might not decay before his body. He assured his friends, that, as far as this might be allowed to depend on his own watchful care over his understanding and his temper, he would preserve himself through the trials of sickness and suffering, to the last, such as they could continue to respect and love.

When he was in his seventy-first year his health began to decline. The first symptom of the infirmities of age—I may say, the only symptom which he had yet felt—was the failure of his eyes, which had been till this late period remarkably good. To him, who made such use of his sight, and who set such value on the power of independent occupation, this was most alarming. He bore the privation with the most cheerful fortitude, not suffering his spirits to sink; sometimes uttering a sigh, but never an unavailing complaint, or one peevish word. On the contrary, he was thankful for the acuteness of his other senses, and with their help, and by the exercise of his ingenuity, found various methods of supplying the deficiency of sight, and of providing for himself occupations and amusements.

In the winter of 1815 we went to Dublin, where he had the best medical advice; he was under the care of Surgeon-general Crampton—a man whose great talents he had always highly admired, and on whose skill in his profession, and kindness as a friend, he had perfect reliance. But all that kindness, and all that skill could do, were tried in vain. The digestive organs were impaired beyond the restoring power of the art of medicine.

From the commencement of this illness, it was my father's opinion, that he might linger a few months, perhaps that his existence might be prolonged a year or two; but that there could be no expectation of recovery, or of any permanent enjoyment of ease or health. This opinion, which was not contradicted by his medical friend, did not diminish the wonderful cheerfulness and activity of my father's mind. His intellectual energy seemed to increase, as his bodily powers weakened; and he pursued his usual occupations with extraordinary zeal.

Previous to this illness, he had been occupied in trying some experiments on wheel-carriages. The same which he had formerly wished to have tried in England, at Woburn. This had been a favorite scientific object of considerable public utility, in which he had persevered during a great number of years, and which he anxiously desired to complete, having always before his conscience the early admonition given, the principle early instilled and followed to the last, to *finish* whatever he began.

He had engaged to try a set of public experiments for the Dublin Society, and with the assistance of his son William he fulfilled his engagement, and accomplished his object. In April 1815, and in May 1816, he tried in Dublin a set of experiments on wheel-carriages, of which an account is before the public*. At this time he went through a degree of fatigue, which no one who knew how ill he was, no one indeed who saw him, could have conceived that his state of health could endure.

---

* In the second edition of his "Essay on Wheel-Carriages."

He returned home, in such a dreadful state of suffering and weakness, as left us little reason for hope. Yet there was so much animation in his manner, such firmness and strength in his voice—we saw him revive so wonderfully in every interval of sickness, and make such exertions in every intermission of pain, that we could not believe the body was failing, while the soul was so full of life and energy. But though his cheerfulness was unabated, and his mental exertions, directed continually to some useful, benevolent purpose, struggled in a surprising manner through sickness and pain; he was well aware of his danger, he never deceived us or himself with false hopes. Nor did he ever for a moment repine, even when, from the loss of sight, he was deprived of all power of independent occupation.

He thanked God, for the blessings still left to him, and with resignation, gentleness and tenderness, most touching in one of his vivacious temper, and independent habits and character, he submitted, as he said, *willingly* to become dependent on the kindness of his friends and children; and he seemed really more sensible of the pleasure of their assistance, than depressed by all the pain and the privations, to which he was subject.

His greatest anxiety during his whole illness was to prevent his wife from wearing herself out, by her unceasing exertions. He expressed a constant fear of being spoiled by her extreme kindness; for he had observed, that men are apt to grow selfish in long sickness, and at last to expect as a right, or to receive without gratitude as a tribute, or a duty, all the continually recurring little efforts of affection and attention. Not a day, scarcely an hour passed, without his giving her some proof of his care in watching over himself, lest he should fall into this error.

"The smallest service merits thanks," he used often to repeat; and his thanks never being mere words of course, or appearing to be uttered as acquittances for kindness, were ever agreeable and touching.

In illness, when many friends anxiously offer their services, it is difficult to the patient to avoid hurting their feelings by refusals or acceptance. My father, however, had the kind art of gratifying his whole family by employing all and each in the manner best suited to their habits and talents, consoling each as far as possible with the feeling of contributing to the alleviation of his sufferings. One of his children wrote, another read for him; one on literary, another on scientific subjects. By one who had followed the course of the experiments in which he was engaged at the commencement of his illness, and who always caught his ideas without giving him the fatigue of explanation, his inventions were still executed with the promptitude he loved. Another of his children, who had habits of punctuality and vigilance, uncommon in one of her early years and vivacious disposition, was, as he loved to call her, "his incomparable little nurse." And with his ever ready, indefatigable secretary, he found, as he said, that without labor to himself, and apparently without effort to her, his thoughts were written almost as quickly, as they could arrange themselves in words.— To her I now owe it, that we possess copies of almost every interesting letter, which he dictated during the last years of his life. From these I have selected such as I trust will, with such notes of explanation as I shall add, give to the public a sufficient view of his mind during his last illness.

TO HIS SON, SNEYD E.

Edgeworth-Town, Tuesday, May 30, 15.

"MY DEAR SNEYD,

"IT gives me sincere pleasure to find, that you and Mrs.

E. return with satisfaction to your own domicile, and that your professional friends seem satisfied, that your absence was occasioned by important business, and not by the temptations of pleasure.

"My friend, Mr. Day, used to say, that nobody was the better for advice either from men or books. Now this might be true of Mr. Day, but it certainly has not been so with respect to me; for I have felt distinctly the instant impulse and the permanent influence of just sentiments, that I have read or heard at every period of my life. For instance, the beautiful poem, 'The Choice of Hercules,' made a strong impression upon me. When I was young, the following lines on Fame caught my attention:

'Who seeks her must the mighty cost sustain,
'And pay the price of fame, labour and care and pain.

"And, as I grow old, these occur to me continually:

'In which no hour flew unimprov'd away,
'In which some generous deed distinguish'd every day.'

"To morrow, my dear son, I hope to complete my seventy-first year—a year exceeding the prescribed term of human life, and nearly two years of age beyond that, to which any of my male lineal ancestors had reached, since our family settled in Ireland. It is true, that my health at present is far from being reestablished, as I cannot, 'by taking food, add one ounce to my gravity;' but my mind is active, and I feel and taste all the various blessings, with which my protracted age has been blest.

'I have liv'd long enough, my way of life
'Is fallen into the sear and yellow leaf;
'And that which should accompany old age,
'As honor, love, obedience, troops of friends,'
I have.

"My sister R——, who is here, has read your Memoirs of the Abbé Edgeworth with interest and approbation; so have others of our friends."

To the same—a few weeks later.

"* * * NOTHING can be more gratifying to a father in the gentle decline of his life, than the gratitude and affection of a well educated son. He feels the wisdom of Providence come home to him with irresistible evidence, and sees that in the order of Nature, where this order is not perverted by the adventitious passions of avarice and ambition, the support given by the parent in the helpless age of childhood is repaid almost in kind, by the assistance and solace afforded to him in his *slippered* decrepitude.

"Nothing solaces my age so much as the unaffected attention, which I receive from all my children, and from none of them more constantly or more effectually, than from yourself.—" * * * *

The observations, in the following extract from a letter to another of his sons, arose from our having reproached my father with being too philosophic in his disdain of ambition.

"I HAVE been frequently told, that my general conversation tends to weaken, and in some degree extinguish, that noble ambition, which arises in every generous mind, not only from the nature of society and the dictates of reason, but which is implanted in our minds instinctively along with life. We see it even in animal life.

"What may be the emulation of fishes, I do not know; but of well-born birds, the quail and the cock—and of well-born beasts, the horse and hound, freely *bestow* in competitions that life, which the inferior of their own species are obliged to *pay*.

"The ambition, which I think unworthy of a noble mind, is a desire for riches, and titles, and stars, and strings, and equipage, and admittance into the company of those, who are engaged in the same miserable scramble. But the distinction to be acquired by high character, height in a profession, and by scientific or literary preeminence, is the great object, and the great incitement to every virtuous and honorable exertion in society."

* * * * * * *

March 7th, 1816.

The following, to the Minister of Louis the Eighteenth, was addressed to him at the time when the public papers announced, that a monument was erecting in France, to the memory of Marie Antoinette and Louis the XVIth.

TO MONS. DE CAZE.

"SIR, *Edgeworth-Town, March,* 1816.

"As it has fallen to your lot, to discover the last writings of the magnanimous Marie Antoinette, and as every heart in Europe is touched by these remains of fallen greatness, I beseech you, to turn, at this moment of sympathy, the attention of your countrymen to the merit of the man, whom Louis the XVIth chose for the depositary of his latest will and wishes, and whom Marie Antoinette calls her friend, and appoints to be her Executor.

"As the nearest relation of the Abbé Edgeworth, I claim from the justice of France, that his name should be inscribed on some public monument, with those of the exalted personages, who relied for consolation on his fidelity and courage. The records of illustrious actions do not perish with brass and marble—they live in the page of history; but even the fragile memorials of art will teach the present and some future generations this rare lesson : That monarchs may have friends, and that princes can be grateful.

"R. L. EDGEWORTH."

The following was addressed to Mrs. Edgeworth's mother, Mrs. Beaufort.

"*Edgeworth-Town, 30th of May*, 1817.

"MY DEAR MOTHER,

"I write from the midst of such a circle of children, as is not I believe to be found any where else. Such excellent principles—such just views of human life and manners—such cultivated understandings—such charming tempers, make a little Paradise about me. To fade away from the world with such objects before me, and at the same time to avoid bodily pain, is all that I can wish.

"We were much obliged to you for your letter* I should defy the most intelligent critic in the world to discover, that it was written by a lady, who had been married for fifty years.

"Maria's tales will soon issue from the press. If they fail of succeeding with the public, you will hear of my hanging myself.

"We heard from Francis (Captain Beaufort) this day, in an excellent and kind letter. I think most highly of his book (Karamania). You know I always speak sincerely.

"Your daughter H—— was the delight of Mary and Charlotte Sneyd, while they were in Dublin.

"Adieu, dear Madam—I was determined you should have one more letter from me.

"R. L. EDGEWORTH."

The tales alluded to in the preceding let-

---

* Written on the 50th anniversary of her wedding-day.

ters are Harrington and Ormond. There may be some impropriety in my publishing what relates to these. But no paltry fear of what may be thought of myself can induce me, to suppress these affecting, and I may say extraordinary proofs of a father's affection for a daughter. After days and nights of sickness and pain, such as would have incapacitated any common mind, my father, in the intervals of ease allowed him, heard every evening, with inconceivable eagerness of interest, what had been written for him every morning; he still pursuing the labor of correction with an acuteness, a perseverance of attention!——of which I cannot bear to think.

(1817) *June 7th.*

Some circumstances recalling to my father's mind the time of the union between England and Ireland, he dictated the following addition to his former remarks on the subject.

"If ever any country was governed by an oligarchy, Ireland was in this situation before the union.

"The only plausible argument, that has been produced in favor of venal boroughs, is, that by their means men of high talents obtain seats in parliament, from which, if they had not friends in the aristocracy, they would necessarily be excluded.

This might for some little time be the case, but the necessity of employing talent would soon be felt, and men of abilities would be chosen for counties instead of being put in for boroughs.

" A contrary system leads to the conversion of politics into a lucrative trade. If it should ever be the case, that men cannot be found to fill the higher offices of government, without being paid as in a profession, the spirit of the English nation will evaporate. Men will then be found to advocate every cause, and to advocate with ability; but love of the country and national glory will be considered as schoolboy, obsolete themes, foreign to the purposes of omnipotent parliament. Some reform in the House of Commons is necessary, and will in time be effected; but this must begin by a reform among the people. The petty electioneering views and expectations of voters are more corrupt, than the designs of the most artful demagogue, that ever obtained a seat in parliament.

" A complete reform in the Irish system of representation actually took place at the union; what its effects may be in checking the influence of the crown, time alone can determine: but there is every reason to suppose, that it will necessarily produce a superior race of representatives, because the representative body must be chosen from men, who have had all the advantages of a liberal education, and all the independence which affluence can confer.

" As I have formerly mentioned, it was at the time of the union in my power, to have followed the career of ambition. I might have obtained a seat in the imperial parliament, and might have dedicated the remainder of my life to what is fondly called the good of my country; or which, under this name, is often meanly pursued, *the advancement of my family and fortune.*

" I now feel most sincerely grateful to Providence, for having given me sufficient prudence and resolution, to resist these temptations, to follow a different course, to cultivate my estate, to improve my tenantry, and to educate my children.

" How far I have been successful in these pursuits, it is

not for me to determine; but there is one enjoyment, of which it is impossible to deprive me; *the consciousness of having endeavoured to do well.*

" I dictate what I am now writing, in my seventy fourth year, in my bed, to which I am nearly confined by a lingering disease; and I would not at this instant exchange the satisfaction which I feel from the attachment of any one of my numerous children, for the largest fortune, or the highest title, that is possessed by the most powerful statesman in the empire."

The following letter was written but five days before he died.—Early in the morning, after a sleepless night, it was spoken with the strength of voice and fluency of one in perfect health. I find, that I had at the time endorsed on it these words: " This letter was dictated to me, just as it is here written, without his ever hesitating or changing a single word."

TO LADY ROMILLY.

" MY DEAR MADAM, *June* 8, 1817.

" I have a strange habit, for which I fear you will not pardon me, of snatching the pen from my daughter Maria, when she is going to write to you. But I will not fill the whole paper with my dotage.

" Formerly it was the fashion, to write under the title of ' Letters from the Dead to the Living.' I think letters from the dying, to those who are in full vigor, might perhaps be full as instructive. But then, who is there that has courage to speak the truth?—Superstition and bigotry press upon the enfeebled mind, and enhance the instinctive terror of death. If a careful retrospect be made of former life; if the faults which we have committed strike us with regret, and if we do sincerely believe that we should avoid them were we to live over again; and if upon the whole we feel the internal conviction, that we have exerted our faculties in the exercise of our domestic duties, and, as far as it has been in our power, for the public good, I do believe, that the descent to the grave, if we can escape bodily pain, may resemble sinking to a sweet sleep. I am now confined to my bed for the greatest part of the day. I suffer considerable pain, and almost constant sickness; and yet my mind retains its natural cheerfulness. I enjoy the charms of literature, the sympathy of friendship, and the unbounded gratitude of my children.

" The manner in which you mention Madame de Staël has occasioned this tirade; and I, with the most true submission of the heart, feel silently grateful to Providence, for permitting my body to die before my mind.

" When I consider what I have just said, I feel that I place myself in rather an awkward situation. What right have I to call up before your Ladyship, who are in the full enjoyment of every happiness the world can bestow, melancholy ideas, which relate to a person, who has very little claim to your attention? And yet I never could consider your Ladyship as a common acquaintance, who took us up from fashion, and would lay us down from caprice.

" We are heartily sorry for Lord L * * * * * * * * *'s illness. I look upon his Lordship to be a first rate *gentleman*; and *I* am inclined to think, that the understanding of the lady is in no way inferior to his.

" The little Dramas, which you mention, are inferior performances, upon which I assure you we set but small value. They however sell well, which we are glad of, on our publisher's account. In a few days I hope you will receive Maria's new Tales. I do acknowledge, that I set a high value upon them. They have cheered the lingering hours of my illness; and they have—I speak literally—given me more hours of pleasure during my confinement, than could well be imagined from the nature of my illness.

"To shew you, my dear Madam, that my mind is perfectly free from gloomy ideas, I send you some lines, which I made the other day in bed*. I have also, within these few hours, dictated some of the anecdotes of my life, in addition to what I had formerly written of the little history of my youth and education. I have also written a little school-book for my son, of which I enclose you a copy, as it is not to be published. It is a most humble performance; but we think it will be useful * * *

"I am, my dear Madam,

"With great respect, your obliged servant,

"R. L. E."

Five days after this letter was written, the writer of it was no more!——His prayer, that he might preserve as long as he lived his intellectual faculties, was granted. To the last they continued clear, vigorous, energetic; and to the last were exerted in doing good, and in fulfilling every duty, public and private.—The strength of his understanding, such as it had been while in perfect health, with more than all that we ever knew of the tenderness of his heart, poured forth with the last efforts of life in his parting words of counsel and consolation to each of his family.

---

* See Appendix.

In his last hours his bodily sufferings subsided; and in the most serene and happy state, he said, before he sunk to that sleep from which he never wakened—

" I die with the soft feeling of gratitude to
" my friends, and submission to the God who
" made me."

He died the 13th of June, 1817.

It remains with his children to do honor to his memory.

The following were my father's directions concerning his burial, left in writing, addressed to his sons, who were his executors.

"I desire, that I may be buried in as private a manner, and at as little expense as possible. I have always endeavored to discountenance the desire, which the people of this country have for expensive funerals. I would have neither velvet, nor plate, nor gilding employed in making my coffin, which I would have carried to the grave, without a hearse, by my own laborers.

"I desire, that no monument or inscription, but one precisely similar to that which I have erected for my father, should be erected for me."

His orders were obeyed. His coffin, without velvet, plate, or gilding, and without a hearse, was carried to the grave by his own laborers. There was the most respectful and profound silence. His remains were deposited in the family vault in the churchyard of Edgeworth's-Town. No monument or inscription, but one precisely similar to that which he had erected for his father, has been erected for him—a plain marble tablet within the church, on one side of the communion table, bearing his name, and the date of his birth and death.

APPENDIX.

Among my father's papers I found the following sketch of a plan for a School at Edgeworth's-Town: it was written many years ago. His son, Lovell, has since carried it, in all its principal points, successfully into execution.

Sketch of a Plan of an Elementary School for the lower Classes in Ireland,

No child should be admitted, that is not healthy:

That is under seven, or at the time of admittance above nine years old.

The children should be taught to read with propriety, and to spell correctly: great care should be taken to make them understand what they read:

They should be taught to write a current, useful, not a fine hand, and should be made expert in the first four rules of arithmetic.

They should be taught as much of universal grammar, as might make particular grammars intelligible, and as should enable them to write and speak English with propriety.

In this school the understanding should be cultivated and exercised, without loading the memory; and the constant object should be, to excite the pupils to think and to apply their understandings to their conduct.

They should be taught the *advantages* of obedience, of truth, of cleanliness, and of sobriety; of sobriety, particularly in this country, where in some places the earliest ambition of a boy is to drink.

A sense of religion, without superstition, should be gradually unfolded in their minds, by books and conversation suited to their tender years.

The ministers of their respective religions should have entire facility afforded them for inculcating their catechism and their religious precepts.

The children should be sent regularly to attend the places of worship, which their parents point out.

A good play-ground, and means of using daily bodily exercise, should be provided, by a due alternation of active and sedentary occupations.

At the age of ten, such boys as are intended for learned professions would, it is to be hoped, quit such a seminary with health uninjured, with habits of obedience, attention, and good conduct, speaking good language, with minds habituated to apply their reason to practical use, and disposed to learn for their own benefit, and not from compulsion.

Such children as were intended for business, or for any active life, might remain at this school, in a higher class, for two or three years longer; and might, in this time, obtain such a knowledge of book-keeping, of the principles of domestic economy, and of trade, as would prevent them from feeling themselves such helpless beings in the common concerns of life, as most young people find that they are, when they are left to provide for themselves.

To this knowledge might be added the first principles of mechanics and natural philosophy, which are universally useful.

Such a school would be of such material benefit to Ireland, that it has for many years been an object of my anxious hopes, to have it established. I think it will be in my power, to accomplish it at Edgeworth's-Town, where some of the requisite advantages already exist, and others I may hope to obtain.

As to physical education, Edgeworth's-Town is a village remarkable for the health and longevity of its inhabitants.—As to religious and moral education, here, happily, the ministers of both the Protestant and Catholic religion are free from bigotry, and live on the best terms with each other, without the slightest mutual jealousy: and as to superintendance, my own family would have a constant eye upon the seminary.—But the difficulty is, to find a proper master.

This elementary school shall be established, whenever a master, and, what is of more consequence, a mistress of the house, can be provided, for whose manners, morals, tenderness, knowledge, and successful experience in teaching, I can dare to pledge myself*.

R. L. E.

Extract from the Appendix to the Third Report of the Commissioners of the Board of Education in Ireland.

TO THE COMMITTEE OF THE BOARD OF EDUCATION, APPOINTED TO REPORT UPON THE CHARTER SCHOOLS.

I congratulate the Board upon the flourishing state of the Charter Schools of Ireland. Beside the satisfaction, which we receive from the prospect of having a number of useful subjects added to the community, we must be gratified by having it in our power, to evince to the Government of the United Kingdom, that the education of children in these Schools is efficacious, practical, free from bigotry, and in every respect such, as to put it beyond the reach of private defamation and public censure. When our Report passes through the hands of Government to the public at large, it will be compared with Mr. Howard's just representation of these Schools at a former period; this comparison will give an irrefragable proof of the gradual and encreasing attention, which is now paid to the

---

* In the school established at Edgeworth's-Town, in 1816, by my brother Lovell, and alluded to in my father's last letter to Lady Romilly, there are now (March, 1820), above 170 boys of the lower, middle, and higher classes, Protestants and Catholics. The seminary flourishes; has succeeded beyond our utmost hopes; and is approved of by both Protestant and Catholic ministers.

lowest classes of people in Ireland. This improvement is owing to the sagacity and perseverance of the Committee of Fifteen, who have wisely entrusted part of the superintendence of the Charter Schools to respectable gentlemen resident in the neighbourhood. By these, and by other means pointed out in our Report, nearly all the Charter Schools in Ireland have been brought to a high state of regularity; the few instances to the contrary, which have been reported to the Board, will, of course, be reformed; and their being reported to us is an internal and indubitable evidence, that the Reports of Dr. Beaufort and Mr. Corneille are not merely an echo of the representations of the Committee of Fifteen, or of Local Committees.

I shall now proceed to suggest to the Committee a few hints for farther improvement of the Protestant Charter Schools; and, first, as to the buildings.

Buildings.]—In most places, infirmaries are wanting; in many, dining-rooms and workshops. To supply these defects, it is necessary, that persons conversant with buildings should be consulted; and it appears to me, that a worthy and ingenious member of our Board, Mr. Whitelaw, might be enabled to furnish proper plans and estimates for these purposes, if he were supplied from the country with rough drawings of the present buildings, and of the ground on which they stand. Wherever any difficulty occurs, it will be necessary to send an architect, to examine the buildings on the spot. It would be superfluous to add, that, wherever I can be of use, my services are at the disposal of the Committee.

The Gentlemen of the Local Committees will in all cases superintend, will provide proper overseers, and will inform the Committee of Fifteen of the local prices of materials and of work.

Building Additions.]—In all additions to buildings, after a plan has been approved of, it is better to employ masons by the day than by the piece; because the junction of the new and old work requires particular care, and this care cannot be expected from workmen engaged by task, and because defects in this part of the business are easily concealed, and cannot be easily rectified.

With respect to infirmaries, it may be observed, that the mode of ventilation should be attended to with care, avoiding the extremes of closeness and heat, on the one hand, and of cold and thorough drafts of wind, on the other.

Proper supplies of water should also be provided, not by ordinary wooden pumps, but by strong iron pumps, such as should not require frequent repairs.

Diet.]—The present dietary has been proved to be excellent, by the best of all possible tests, the health and strength of the children. It has been said, that they uniformly prefer potatoes to wheaten bread; perhaps quantity in this case compensates for quality. With respect to *stirabout**, there is reason to believe, that food, which passes down the throat so readily, is neither so agreeable nor so wholesome, as that which undergoes sufficient mastication: It has also been said, that the *stirabout* has been laid aside in many places, because it promotes cutaneous diseases; for this however there does not appear to be sufficient foundation; for our Report states, that oaten bread is used in some of these Schools, and at the same time it is stated, that even in these but few boys were infected with any cutaneous disease. It might, however, become a useful subject of experiment and enquiry, which it is in the power of the Board to prosecute with very little trouble, and with great and permanent advantage to the public.

Religious Instruction.]—This in almost every School fully answers, and sometimes surpasses, expectation; the catechists most laudably attend to their duty, and their labours are successful. The Tract called "*The Protestant Catechism*" had been omitted in many places; it is now entirely discarded. After all that has been said by others, I shall, in as few words as possible, express my own sentiments; it is my duty to do so, or I should decline the subject altogether. The highest authority, that public station and private character can create, has sanctioned

---

* Porridge made of oatmeal and water.

the opinion, "That whatever a good parent of the higher "ranks should do for the religious instruction of his own chil-"dren, should be done for the poor." This benevolent and pastoral sentiment I am reluctantly obliged to question: Children of opulent parents have their minds cultivated by various knowledge; they have abundant sources of instruction from books and conversation; they are thus taught to discriminate, and even at an early age to reason. At a Charter School, the children are with great propriety kept separate from society, and no books get into their hands, but such as their masters chuse to give them; they should therefore be taught dogmatically. The doctrines of our church should most certainly be early impressed on their memories, and they should be made acquainted with the nature and tendency of those errors, against which we protest; but I would by no means prepare them to be disputants, were they capable of entering the lists; I should fear, that they might burst from the hive a swarm of sectaries.

The absurdities of Popery are so glaring, "that to be hated, "they need but to be seen." But for the peace and prosperity of this country, the misguided Catholic should not be rendered odious, he should rather be pointed out as an object of compassion; his ignorance should not be imputed to him as a crime, nor should it be presupposed, that his life cannot be in the right, whose tenets are erroneous. "Thank God! that I am a Protestant," should be a mental thanksgiving, not a public taunt.

General Instruction.]—Writing, reading, and arithmetic, are the standing objects of attention. Of these, *writing* is in general well taught; reading, not quite so well; and arithmetic less generally, and perhaps less successfully, than might be expected: yet of all the common acquirements, of which the young mind is capable, arithmetic is the most useful; its rules are logical, their foundation is laid in immutable truth, their development excites and gratifies early curiosity, and it is impossible to have learned the higher rules of arithmetic under a good master, without having the powers of the mind improved; and what end can be proposed more advantageous to society in the education of the poor, than to give them good sense, and reasoning minds? To make the poor tractable, you must give them sufficient powers of discernment, to discern their real interests amidst the sophistry of those, who endeavour to mislead them.

Books.]—To form the judgment and influence the feelings of the children, beside the instruction of their masters, proper books must be employed. I have been told, that in some schools the Greek and Roman histories are forbidden; such abridgments of these histories, as I have seen, are certainly improper; they inculcate democracy, and a foolish hankering after undefined liberty: this is peculiarly dangerous in Ireland. Among the many books, which may be advantageously permitted, I shall presume to mention the following: "Barbauld's" (beautiful) "Hymns," "Moral Annals," and "Butler's Arithmetic," which is full of solid useful facts, adapted to every pursuit of their future lives; also "Butler's Geography," with any other compendium of Geography, that mentions the products of different countries.

It is often said in England, that an Irishman does not know his right hand from his left. Let our poor children be taught at least the cardinal points of the compass; let them learn to know the pole star, and three or four of the constellations, the causes of day and night, and the annual motion of the earth; even Caliban is proud of these acquirements. The principles of draining, and similar parts of agricultural knowledge, applicable to the situation of the lower classes of the people, may be advantageously taught.

The children should see specimens of the common poisonous plants and minerals, and antidotes should be pointed out to them.

For their amusement, stories inculcating piety, and morality, and industry, should be admitted. But every thing that leads to restlessness and adventure should be carefully avoided. The attention should be turned as much as possible to sober realities. For instance, the habit of estimating measurement should be early taught, it enlarges and occupies the mind, and is of daily use in every situation of life. A competent portion of what is here mentioned might be taught by masters visiting these schools from time to time, without much trouble or expense.

Employment.]—To find proper employment for children is a *desideratum* not yet attained, but it may be approached. It is always in the power of the master, to encourage gardening; it is profitable to him, and healthful to his pupils: nurseries of trees are still better sources of employment and of profit, than common gardening*. Boys take an interest in what at the same time occupies their minds, and employs their limbs; this interest will grow up with them, and may by degrees supplant that hatred for trees, or that love of destroying them, which, it is said, subsists in Ireland.

Knitting and spinning are totally unfit and unprofitable for boys; weaving is a healthful exercise, if not followed with too much assiduity; the flax mills now establishing in Ireland will soon supply materials every where. Is it to be supposed, that the legislature will refuse to supply looms?

The boys should not work more than three hours a day. Looms for cotton and woollen goods should of course be employed in some places, instead of those for weaving narrow sacking and coarse cloth.

Stocking weaving instead of knitting should be introduced; netting, and weaving sash-cord, curtain line, and fringe for furniture, might be tried. Basket-making is a good employment; shoe-making is already taught, and it may be more generally introduced, for shoes are becoming every day more common in Ireland. The hours for play in the schools are not sufficient; ball-playing, gough, and cricket, and all manly sports should be encouraged: "Mens sana in corpore sano" is the description of a useful citizen. A book should be kept, stating privately the genius, merits, faults, and progress of every boy in every school: from each a certain number should be selected every year. And different schools should be established, either upon

* *Note by the Editor.*—A nursery of trees has, within these last two years, been made for and by the school of Edgeworth's-Town, and has afforded constant useful employment.

the present, or upon a new foundation, to breed boys to different occupations; servants, shoemakers, cabinet-makers, clerks, merchants, surveyors, schoolmasters, parish clerks and choristers, and soldiers, who must soon from their acquirements become sergeants, and might then by their education be brought forward in society. It was thus, that the Jesuits made their pupils superior to those in any other seminary on the continent. "Fas est et ab hoste doceri."

In all cases the work-boy and his master should have part of the profit of their industry; and surely the master should be encouraged to look forward, as he grows old, to a permanent establishment for life.

Consolidation of Schools.]—Upon the whole, the greatest improvement, that could be made in these schools, would be to reduce their number; four schools, containing from six to seven hundred, according to local circumstances, would answer all the ends proposed by the present diffused establishments. The superintendence of these schools would be easy and effectual.

That discipline by which armies are governed, which cannot take place in the management of a few boys, might be introduced amongst numbers; the division into small bodies, with the system of gradual subordination, and promotion in consequence of merit, would induce habits of submission and emulation, which would be carried from the school into every situation, where the boys might afterwards be placed.

It would be practicable to send a master in rotation to these schools for a fortnight, twice a year, to teach various useful parts of knowledge, some of which are before mentioned. From the impression made by incidental instruction, the bent of each boy's disposition might be learned, and his proper destination might be ascertained. A useful and cheap apparatus for teaching the principles of natural philosophy might be had for this purpose, and a proper master be found, who should not aim at teaching more than what is obviously useful. Were this effected, apprentices so educated would soon be in such high request, as to make it an object of competition amongst the parents of the poor, to have their children admitted into

Charter Schools; and then by degrees the foolish prejudice against this mode of (charter-school) education would be eradicated, a circumstance which might in itself be of very high advantage to Ireland.

Nov. 8, 1808. RICHARD LOVELL EDGEWORTH.

From the Appendix to the Fourteenth Report of the Commissioners of the Board of Education in Ireland.

LETTER FROM RICHARD LOVELL EDGEWORTH, ESQ. TO HIS GRACE THE LORD PRIMATE*.

MY LORD,

In obedience to the Resolutions of the Board, I offer to your Grace the result of my reflections on the Education of the Poor of Ireland. I am thoroughly sensible of the importance, and as fully sensible of the difficulty, of the undertaking.

The public expectation has been much excited by the appointment of this Board, by the respectability of its members, and by the prudent silence, which it has hitherto preserved. Enthusiasts imagine, that some extraordinary scheme may be devised, which shall at once change the views and the habits of the population of Ireland; while others, aware of the difficulties which occur on every side, despair of our being able to effect any material improvement in the present modes of education, and smile at those, who turn their attention to such a hopeless inquiry. It should, however, be recollected, that within half a century a prodigious improvement in the manners and habits of this country has taken place; and it seems evident, that this improvement has arisen from the better education of every class of its people: there are more Schools, there are better books; and the private advantage of some degree of literature is more generally understood than formerly. For one person that could read or write twenty years ago, there are now twenty; and the same advancement in every species of knowledge may be perceived in every city, and every village in Ireland.

* Hon. Dr. Wm. Stuart.

If this has been the case, under the present modes of Instruction, which are obviously defective, is there not just reason to suppose, that a more rapid and extended benefit may be obtained by a better system?

There are, I am well aware, persons who altogether deny, that any improvement in the people has resulted from their having had more education. There are persons, who oppose instructing the poor even in the elements of literature, because, say they, if the poor are taught to read, they may read what is hurtful; on the same principle we might as well object to the appetite for food, because poison may be swallowed instead of wholesome nutriment. That pernicious books are now read in the present day-schools of this country, is certain; but this arises from the negligence of those, who superintend these Schools, and who do not put proper books into the hands of the children. Does any rational being imagine, that there is an innate or unconquerable propensity in the human mind for reading *only* the "Spanish Rogue," or "the Adventures of Captain Frene?" Put good books (I do not mean *merely* religious books), that shall entertain and instruct them, into the hands of the children of the poor, and they will soon form a taste, that must disdain such disgusting trash. To prevent the circulation of what is hurtful, the utmost care should be taken in selecting books for schools, and none should be introduced without the sanction of those, by whom the Masters are chosen. In the mean time it must always be a sufficient answer to those, who object to teaching the rudiments of knowledge to the poor, that of three thousand boys, who have been educated at the Sunday schools in Gloucester, only ONE has been convicted of a crime: that of four thousand educated at Lancaster's schools, not one has ever been brought into a court of justice; and that the humane and observing Akerman, who was some time ago keeper of Newgate, asserted, that not one person in a hundred of the prisoners, who had learned to read and write, had been executed during the time that he had been governor of that prison.

To attempt to controvert such facts by declamation, by assertion, or by the mere opinion of individuals, is merely beating the wind. It is true, and it is fresh in our memories, that, in the progress of the last rebellion in this country, those who could read and write were, at first, employed to inflame and direct the rest; but there is good reason to suppose, that this happened from there being but few, that could by writing carry on the schemes of the disaffected; and that the preeminence and temporary consequence, which these fellows acquired from their being able to read and write, was the cause of their being more easily induced to disaffection: this would not have happened, could numbers have been found, who had the same means of becoming useful to their leaders; and we may further observe, that these *scholars* were then taught a useful lesson, which cannot be easily forgotten; they found, that, when brought into action, more ignorant and more desperate men took the lead, and the scholars felt, that they were neglected and despised.

It has been said, that a Gentleman, to whom the world is much indebted for a large share in the late improved methods of teaching, I mean Dr. Bell, has given it as his opinion, that arithmetic is not a necessary part of early education. I am obliged to declare an opinion, that is precisely contrary to this —if it were necessary to dispense either with reading and writing, or with arithmetic, I should rather dispense with the two former, than with the latter. I think it was Swift, who, when he was asked what the Irish nation needed most for its improvement, replied, "to learn that two and two are four."

I consider arithmetic as the most instructive science, that can be taught to children. It is the first occupation of the youthful mind, that disciplines it to think with accuracy; and whoever has learned the common relations of numbers, whether he has learned by the eye, or by the ear, has made an advance in accurate reasoning, that cannot be so easily or so certainly attained by any other process, that has been yet discovered. The names of numbers, and the figures which denote them, are symbols by which a perfect system of induction is carried on by the understanding; and whoever has once acquired a clear notion of this mode of reasoning may advance gradually to the most difficult problems in every human science. Arithmetic is not only the most certainly useful, but the most securely safe, acquirement for the lower ranks of the people; from books, if ill chosen, they may learn error; from the relations of numbers nothing can be extracted but truth. I do not know on what grounds arithmetic can be objected to, except on the truly Popish principle, disavowed, indeed, by all liberal Catholics, that the people should not be taught to *think*. This principle is as dangerous, as it is illiberal; for it is in these days absolutely impossible, to prevent the people from thinking.

The progress of knowledge has spread now so far, that it cannot be stopped without destruction to those, who attempt to arrest its course. The people *will* read, and *will* think; the only question that remains for their governors is, how to lead them to read such books, as shall accustom them *to think justly*, and thus make them peaceable subjects, and good members of society.

We must next examine, whether the difference of religious creeds, and the animosities of party prejudice, can be so far reconciled, as to permit the adoption of any general system for the instruction of the people. It is not intended, that Protestant masters shall interfere with the religious instruction of Catholic children; and it is still more vain to suppose, that, among a number of Catholic masters none could be found, who would endeavour to teach what they believe to be salutary truth to the children of Protestants committed to their care.

There is but one method, that appears to me practicable in this state of things: to let Protestants appoint masters for Protestant children, and Catholics choose masters for their own schools. The obvious objection, that arises against this scheme, is, that it draws a line of demarcation between the two sects, even during childhood,—that it separates Catholics and Protestants;

and that it avows a deep suspicion and jealousy, which ought not to exist between members of the same society. But this theoretical objection must give way, when we consider, that this separation can last *but a few hours* daily; that these very Children will converse and play together promiscuously; and that this temporary separation must prevent comparisons and jealousies, that naturally arise where contending sentiments and contending interests may be exposed to collision. In another point of view, this temporary separation, far from tending to estrange the sects, will, by showing distinctly that there can be no scheme to undermine the speculative opinions, create confidence among the parents and clergy of the Catholics. The clergy, and in particular the superior clergy, will find themselves treated with the consideration, which is due to ministers of the Gospel, whatever may be their particular creeds, where their lives are not in contradiction to their professions; nor is this mere *ostensible liberality*, a word that has been of late degraded; but it is *fair-dealing;* an expression somewhat more homely, but not less significant.

There are many places in Ireland, where Protestants and Catholics are taught to read and write, and to say their respective catechisms, by Catholic masters,—there are on the contrary other places, where every attempt of the most enlightened and benevolent people has failed, to collect the children of Catholics under a Protestant master; but in most places it has been observed, that, where no particular circumstance has arisen to awaken religious animosity, or well-founded suspicion, the best teacher, whether Catholic or Protestant, soon attracts all the scholars, and the inferior master is obliged to give way; and it is obvious, that in all cases, ***where the two sects agree, there need be no separation.***

There are persons who think, that the allowing Catholic bishops, or Catholic clergy, to have any share in the superintendence of schools is unsafe, and that it is a degradation of the dignity of Protestant clergy, to act along with them. How this opinion can be made consistent with the clerical character, or with that Christian charity, for which all the ministers of the Gospel ought to be distinguished, I cannot imagine: of this I am certain, however, that such an opinion can never in a political point of view be safe or prudent. It can never be good policy, to degrade the ministers of the Catholic religion in the eyes of the people, whose consciences they are to direct, and whose morals they are to form.

Having now fairly stated the principles, on which I would found any attempt to improve the national education of this country, before I further explain my plan, before I suggest any thing new, I think it necessary to say, that I would not undo any thing, that has been already done; that I would not, for the chance of making it better, destroy any good, that actually exists.—As the enlightened and eloquent Burke observed, those are rash and ignorant reformers, who begin by the destruction of existing establishments, especially of those intended for the education of youth. In such establishments, which time and custom have consolidated, even though they may not be the most perfect of the kind, yet there is always to be found a *power*, what the workmen call a *purchase*, of which the skilful legislator can avail himself, and which he can apply to useful purposes.

Far from wishing to destroy what has been already done far from wishing to abolish the parish schools, I am thoroughly convinced of their utility; and I hope, that the bishops of the established church will exert their just authority with respect to these schools; and that the stipend payable by the incumbent of the parish should be raised to at least six-pence in the pound upon the clear income of his living,—that a general fund should be made of these contributions, so that it may be afterwards appropriated to the wants of different parish schools. Each parish should be obliged to keep in repair such school-houses as have been already built, and should be obliged to build, where schools are by the present laws appointed to be kept. The regulation of these schools, I think, should be left entirely in the hands of the clergy, by whom they are supported; upon their prudence and good sense the people must depend for their being administered with liberality.

Beside these parish-schools, I propose that a number of new schools should be established. These I would divide into two classes, preparatory and provincial. From the preparatory schools, which should be day-schools, I would have a certain number of boys selected from time to time, and draughted into the provincial schools, where they should be clothed, lodged, boarded, and instructed, for two, or perhaps three years, at the public expense.

Thus a considerable number of boys, of the best conduct, and of the best abilities, would be taken from the ranks of the profligate and ignorant, and would be indissolubly attached to the laws and government of the country. I would begin upon a small scale, and would feel my way through the obscurity and difficulties, in which the subject seems to be involved. In the first place, I propose, that proper slated houses should be built for these schools, under an intelligent inspector, who should take substantial security for the execution of the work, and for its being kept in perfect repair for twenty years.

Of the preparatory schools, I propose that about thirteen hundred should be established; that is to say, about forty for each county in Ireland; and sixteen provincial schools, four for each province. These should be erected, not at precise distances from each other, but in such places as should suit the population of different districts.

The masters of these two classes should be Protestant or Catholic, according to the prevailing religion of the place where they are built. The greatest care should be taken in the choice of these masters, and they should be removed immediately upon a report of ill conduct made by the inspectors, or upon such information as the commissioners could rely on.

With respect to the emoluments of the masters, the masters of the subordinate schools should be paid partly by a fixed salary of 20*l*. a year, beside a house and garden, and partly by the parents and friends of the children, who are committed to their care; so that, on the whole, the masters of these schools might earn from forty to sixty pounds a year. It is to be supposed, that the payment by the scholars to each schoolmaster will amount at least to twenty pounds a year more: different prices will be proper for different places: in some places, the poor are scarcely able to afford any payment; but it is obvious, that the best charity the higher ranks can bestow is in such cases paying the small stipend required for the schooling of children; nor is there any danger, that this charity should become onerous, as the poor are averse from receiving *gratuitous* instruction.

The salaries of the masters in the higher or provincial schools should be at least one hundred pounds a year, with a prospect of a pension of twenty or thirty pounds a year, according to their respective merits, if they choose to retire after twenty years service. The masters for both classes of these schools should be appointed by Commissioners, to whom the control of these schools should be ultimately confided.

Whether these Commissioners should receive any emolument for their trouble, is a question that I am not competent to decide; but it appears to be reasonable, that either all or none should be paid. The danger of creating Commissioners with salaries is the opportunity for Parliamentary *jobbing*,—the danger of gratuitous superintendence is *neglect:* these Commissioners should be chiefly laymen, half their number should consist of Protestant, and half of Catholic gentlemen: the Protestant part of the Board should choose the Protestant masters, the Catholic part of the Board should choose the Catholic masters; but neither Protestant nor Catholic masters should be chosen, without a certificate of good behaviour from the Protestant or Catholic bishop; or from the resident Protestant or Catholic clergyman; under these Commissioners inspectors of all these schools should be appointed, two for each province; these inspectors should be handsomely paid for each visitation, upon which they should be sent: they should keep regular books, and should report regularly to the Board; their visits to the preparatory schools should not be at stated, but at unforeseen times; and their general business should be to examine and decide upon the merits of the boys, who are to be draughted from the preparatory into the provincial schools.

To accomplish the purposes of these establishments, the boys in the higher schools should be taught book-keeping, surveying, agricultural economy, practical mechanics, and such parts of practical chemistry, as are useful in the trades and occupations for which they are designed.

It may be supposed, that in each of the thirteen hundred preparatory schools from forty to eighty boys may be taught reading, writing, and arithmetic; the masters should be obliged to keep a weekly register of the morals and acquirements of every boy in the upper class of each school: this register should be kept by simple marks, under the heads of truth, honesty, obedience, and scholarship. The inspectors should verify the contents of these registers from time to time by inquiry, and by examination of the boys in the different branches of their instruction.

Once in every two years the master of each school should select two boys, in conformity with the evidence of his registry, to be sent to the nearest provincial school as a candidate for admission; and the boy should be previously furnished with a certificate of good behaviour from the Protestant or Catholic clergyman of the parish, where the preparatory school is situate; and once in every two years, at each provincial school, a public examination should be held of the candidates for admission, before two of the inspectors already mentioned: thus forty boys would be draughted into each of the provincial schools, while at the same time forty would go out to different useful and profitable occupations. If the boys were well taught, they would be eagerly sought for by persons, who were engaged in the employments for which the pupil is prepared. The total expense of this establishment, including the money spent in building, would amount to about 80,000*l.* per annum.

From the structure of this plan it is obvious, that half the number of the provincial schools might be established, or any smaller number, as an experiment: this would diminish the expense to 40,000*l.*, a sum which might be raised by county presentments.

A distinguished member of our Board has observed, that many of the evils, which we suppose to arise from want of education, or from difference of religion, in Ireland, arise from difference of language*, from the lower classes continuing to speak Irish, instead of learning English. This may be the case in some parts of the country, but certainly not in the county where I reside; wherever it is the case, proper methods should be taken for remedying it; the multiplying the number of English schools seems to be one of the means most likely to succeed. It should be considered, for the honour of the docility of the Irish, that they have within these few years made a greater progress in learning English, than the Welsh have made since the time of Edward the First in acquiring that language.

It would be superfluous to enter into minute details upon the plan, which I have laid before your Grace, as it is offered for the consideration of superior wisdom; but I may be allowed to add a few explanatory hints on the mode, by which the simple and easy course of instruction I have recommended may be carried into execution. In the preparatory schools for teaching reading, writing, and arithmetic, advantage should be taken of all the improvements, which Dr. Bell, Mr. Lancaster, and others, have suggested; and their plans may be still further improved: there are means of teaching children to read with more ease, more certainty, in much less time, and at less expense, than any that are in use at present at public schools; but no particular mode of teaching should be exclusively enjoined; the best will soon make its way by its own superiority.

After the second year of the establishment of the preparatory schools, the boys should be divided into two classes, an upper and a lower; the second class should be taught by monitors chosen from the first class, but I totally disapprove of the indiscriminate appointment of monitors; great care must be taken in their selection; only the best informed, and the best tempered boys should be employed; good temper should be preferred to abilities, because, in teaching, good temper is of more consequence than the most shining abilities.

After the second class has been unremittingly employed for about two or three hours, it should be entirely dismissed, and the upper class should remain, and should be taught what might be suited to the age of the pupils, and to their previous acquirements.

Wherever girls are taught, they should be dismissed with the younger class.

It will be immediately objected, that the time I have mentioned cannot be sufficient for any profitable instruction; and that one of the great advantages of a day-school arises from its keeping children employed, and consequently out of mischief, for the greatest part of every day during the year, holidays excepted.

To the former of these objections I answer, that long and attentive experience has convinced me, that much less than one hour's lively attention in the pupil will improve his understanding, under proper teachers, more than ten hours listless impatience under the tuition of a common pedagogue in a common school. As to the second objection, it is best answered by observing, that one of the inconveniences, of which the poor complain, in respect to the education of their families, is, that they lose the assistance of their children, which, though of no very great amount, is yet an object to them in their day's labour; the advantage of schools, as repositories, or rather prisons, for active children, who are troublesome at home, must surely relate to children of very tender years, who ought not to be admitted into the public schools.

The examples that young children see at home are undoubtedly pernicious, but till a better educated generation has grown up, there is no remedy, except what may be expected from the interference of the gentry, particularly the ladies of

Ireland, who are now intent upon bettering the condition of the poor; by their means, *Dame Schools* may be provided as receptacles for young children, to habituate them to cleanliness, order, and obedience, before they are sent to any of the preparatory day-schools, which I have described.

Whatever plan may be adopted for the education of the lower classes, a seminary for masters is indispensably necessary; some of the most promising pupils from Dr. Bell's and Mr. Lancaster's schools might be invited to this country; a succession of persons properly qualified to be masters might afterwards be supplied by selections from our own schools. By proper encouragement, I think a school for masters might be established at Wilson's Hospital. I cannot quit this subject without observing, that the poor are now uncommonly anxious to procure education for their children: as a proof of this, I may mention, that in a number of private letters, which I have lately had an opportunity of seeing from young men abroad in different parts of the world, I have found most urgent entreaties to their parents, or their wives, *to keep their children to school;* this anxious desire, that the children should be instructed, is the best preparation, the best omen, for the success of a plan of popular education; and the plan I now propose would hold out many peculiarly alluring circumstances: the keeping of registers in the schools; the selecting, from the evidence of these registers, the most deserving pupils, without distinction of religion, to be sent to public examinations in the provincial schools, would, in the first instance, give confidence in the impartiality of the system, and excite strong emulation; the further certainty, that the successful candidates at these examinations would be sent to the provincial schools, where, without expense to the parents, their education would be continued so as completely to prepare them, at their entrance into life, for employments and *situations* in a rank or step above their own, must operate as a powerful motive, both on parents and children; a motive which would excite the energy of the young, and secure the cooperation of the old: the poor would see that advancement in many lucrative and honourable occupations is thus laid open to industry and merit; they would perceive, that those only enjoy rational freedom, who have thus the power of obtaining, by their own exertions, what, in other countries, is reserved exclusively for persons, who are born in the higher ranks of life. The riches and distinctions, that may be acquired in many occupations, will thus be considered as a fund opened to every individual in the state; and though, in human affairs, a multitude of unforeseen circumstances retard and obstruct the advancement of individuals, yet where the way is open to all, none can justly complain of being necessarily kept down below their fellow citizens.

Whilst I write thus of what *may be done*, I do not mean to assert, that what *may be done* will be soon accomplished. A generation must pass away, before the advantages of a good system of national education can be generally perceived in the improved morals and manners of the people.

But your Grace's views, and the views of this Board, are not confined only to the present generation: If a solid foundation be laid by your exertions, time will mature what shall have been begun, and the blessings of good education will increase the security and happiness of Ireland, beyond the most sanguine hopes of that government, which instituted your Board.

I cannot conclude this sketch better than with the striking sentiment expressed by the late bishop of Elphin, in his sermon before the Incorporated Society:

" Education makes all the difference between wild beasts and " useful animals, all the distinction between the Hottentot and " the European, between the savage and the man."

I have the honor to be,

Your Grace's most humble servant,

RICH. LOVELL EDGEWORTH.

Extract from the Second Report of the Commissioners appointed to inquire into the Nature and Extent of the Bogs in Ireland, page 179. *On District, No.* 7.

I SHALL proceed to point out the means, which I would pursue, in attempting to reclaim a large portion of Bog.

I would in the first place cut off all the springs and sloughs by appropriate drains; and then, whenever a Bog is covered with thin heath, or weak plants of any other sort, I should endeavour in the first place to burn off the rough surface in a dry summer, merely by setting fire to the heath, and permitting the fire to spread itself, wherever it could meet with fuel. In other Bogs, where the roots of plants are strong and deeply sunk under the surface, I would immediately proceed to turn up the whole of the Bog with a long loy, a tool which is common in the western part of Ireland; besides which I should require grubbing-tools.

As I advanced, I would cut narrow drains of eighteen inches deep, with narrow Essex draining-tools, wherever water lodged on the surface; and drains of fit dimensions, wherever springs occurred, which I had not previously discovered.

The thick tough surface of the Bog should be piled up in ridges to dry, and where the turf so dried yielded ashes that contained manure, as much of it as is possible should be burned in the rows as they stand, to save the expense of making it into heaps; what remains unburned after this attempt should be collected in heaps, and burned to ashes. The next operation is to put out some kind of clay upon the ashes, after they have been spread, taking care soon to cover them with earth, to prevent them from being blown away.

I propose for this purpose to employ wooden portable rail-ways, shod with iron; upon these portable rail-ways appropriate carriages are to be employed; these are made to empty at either side of the rail-way, and when loaded with half a ton of earth, one of them can be easily pushed forward by one man.

I had the honor of showing these rail-ways to several of my friends, in 1787, at the house of Mr. Foster, who was then Speaker of the House of Commons of Ireland. The advantages proposed by this construction were to extend the use of rail-ways to temporary purposes, and to reduce the expenses considerably, by dividing into several parts the weight which is usually loaded on one carriage, so that the rail-way might be reduced in weight and cost, without being more liable to break, than those that are in common use.

This plan has since been successfully adopted in many places, particularly at Penrhyn slate quarry, where a number of small carriages, loaded lightly, supply the place of one larger carriage; by these means, iron rail-ways may be made sufficiently strong, though not one fourth part of the weight or cost of ordinary rail-ways.

I propose, that these portable rail-ways should be supported on piles of five or six feet long, driven into the Bog; and that they should be removed from place to place: so that after a series of rail-ways had been laid. to the distance of half a mile, for instance; and after the carriage had emptied its load on each side of it, the rail-way may be detached, and placed at the distance of two perches, parallel to its former situation; these rail-ways are described, and an estimate given of their cost, in Appendix, No. 4*. It will thence appear, that the share per acre of the capital of 500*l.*, expended on the rail-ways and carriages, would, on 1,200 acres, amount to about five shillings per acre; wear and tear about three shillings more; spreading, gravelling, and removing machines, about ten shillings. This estimate is made on the supposition, that the machines and rail-ways would last four or five years, the time necessary for completing the improvement of 1,200 acres, so as to make the land worth thirty shillings per acre; the expense of the whole per acre would be nearly as follows:

*** To the Reports of the Commissioners for the Improvement of the Bogs of Ireland.**

	£.	s.	d.
Wear and tear, and share of capital, per acre -	—	8	—
Draining - - - -	—	10	—
Turning up surface - - -	2	13	—
Digging and filling clay, &c. - -	2	13	—
Carrying out clay on movable rail-ways -	1	14	—
Spreading clay and shifting the rail-way -	—	10	—
Damages for Gravel-pit - - -	—	7	—
Irish	£8	15	—

(Equal to *5l.* English per English acre.)

To improve the bogs of the whole district, viz. 21,367 Irish acres, would, according to this estimate, cost 181,619 Irish, equal to 167,648*l.* English, for 34,569 English acres; thus a permanent income of nearly thirty thousand pounds English could be obtained at less than six years purchase.

I consider the plan that is here recommended as the first stage of improvement; and I believe, that it would make the Bog worth five and twenty or thirty shillings per acre.

Substantial and intelligent farmers would for half as much more double its value; poor tenants, if they are allowed to have more than a garden, and as much land as will support a cow with grass and hay, would soon wear out what had been done. Reclaimed Bog must be continually attended to; and if red ashes are to be found below the surface, and lime in any form be within reach, the Bog may be made worth four pounds per acre.

I have stated, that the first stage of improvement of a Bog, which is in fact by far the most difficult of any other, may be accomplished for 8*l.* 15*s.* Irish per Irish acre, equal to *5l.* English per English acre; I am well aware, that this appears to be a very low charge, and not unlike those estimates, which sanguine or interested engineers hold out, to induce the public to pursue some favorite scheme of the projector: but what is here laid down is taken from the common prices of work, and from repeated experience of the machinery that is to be employed.

As I never was more thoroughly convinced of the feasibility of any project, than of that upon which I now offer my opinion, I hold myself ready to undertake the reclaiming one thousand Irish acres of a Bog of middling quality; that is to say, between what is most easy, and what is most difficult in this district, for eleven thousand six hundred pounds Irish, which is one third more than the amount of the foregoing estimate; at the same rate 1,000 English acres would cost 7,800*l*. English. If a considerable number of acres of Bog were thus undertaken by persons, who could give good security for the performance of their contracts, the improvement of the Bogs of Ireland would of course gradually follow.

Many Bogs, after they have been drained to a certain degree, are capable of being planted without any farther care, than to dig up and break the surface round each plant; wherever water settles after the trees are planted, it must of course be drawn off. One of the surest indications of the fitness of any Bog for a plantation is the flourishing appearance of the plants already growing in or near the Bog. The growth of bog-myrtle, and of some trees that had been planted on the borders of Bog, made me point it out (in Appendix, No. 5*) as a proper place for an experiment upon planting Bog without first claying and breaking it up. Among the young trees, that I saw thriving on this Bog, I was surprised to find some vigorous larches; I have since seen in Bogs, in the Queen's County, other instances of thriving larches.

In planting Bog, attention should be paid to the probability of a demand for the timber. Near the coast all kinds, and any quantity of timber, may be advantageously disposed of; in this district nothing but oak, ash, and fir, would be profitable.

It appears to me, that Bogs apparently alike are not always equally proper for the same kinds of trees: it would therefore

* To Reports of Commissioners.

be prudent, to make previous trials, before the large plantations should be undertaken. Most Bogs may be prepared for trees by culture; I should therefore recommend, that plantations for shelter and ornament should be made on every Bog that is improved; a judicious assortment of timber trees on these Bogs would not only be profitable, but would entirely change the face of the country. If large plantations were made, they should be toward the middle of the Bogs, lest they should prevent that free current of air, that is necessary to keep the ground sufficiently dry. One advantage attends the plantation of Bog, that is not to be met with elsewhere: the drains, that are necessary in a Bog, may be so disposed, as to form inclosures for trees without any additional expense. It will not be difficult for a person of taste, who is at the same time a skilful planter, to lay out Bogs so as to become highly ornamental, without producing more shade than is wanting.

In the choice of trees, experience alone should be the guide; in some Bogs, as in some soils, one species of tree will not grow, whilst another flourishes in full vigour. I have seen oak flourish well, where ash was wretchedly stunted: perhaps oak may succeed in shallow Bogs, when the plant has strength enough to force its roots downward through eight or ten feet of Bog to solid earth below. I do not however recollect, to have seen oaks growing far from the edges of a Bog; and I have been told, by a gentleman of observation, that the roots of no trees can live in the lower part of deep Bogs. My own experience on this subject is but limited. I have seen large and strong Scotch firs in deep Bogs, and some good alders. Aquatic trees, it is said, have the property of drying wet ground, where they have taken root; if this be the case, they might be employed as *preparatory crops*. Timber sally (salix viminalis) is commonly found in Bogs; whether it will thrive in them in their present state, I do not know; but I have reason to believe, that common osiers do not grow well in such soil.

The other purposes, to which reclaimed Bog may be applied, are so various, and so much the same as those for which other land may be used, that it is not necessary in the present stage of the business, to enter into any minute detail upon the subject. The rich will at first plant ruta-baga and hemp ; the poor, potatoes and flax. And according to the views of the Legislature to encourage the poor to cultivate hemp, the most certain method is to engage for the purchase, at a fixed price, for all hemp raised within a given time. Let the price be liberal, and payable at the principal towns in the neighbourhood, without affidavits or certificates. Some slight frauds may be attempted ; but encouragement for perjury will be thus avoided ; and even the few attempts to deceive, that may be practised in the beginning, will be discouraged by the fair dealer. The peasant, from his own experience, by taking into account the loss of time and credit, will by degrees learn to conduct himself more wisely and more honestly.

The more remote employment of reclaimed Bog should not be unnoticed ; it is a common belief, that after some time it returns to its former state. This opinion I have heard most frequently from persons, who were at the time feeding cattle upon reclaimed Bog ; these unconscionable people, after taking two crops of potatoes and two of oats from the new land, sowed it with coarse grass-seed ; mowed it till it no longer yielded meadow ; and then complained, that it was throwing money away to improve Bog !

In most of the cases that I have examined, I have found, that the capital laid out has been accurately remembered, but that the returns made by the first crops were forgotten ; so that only the present produce of the worn-out soil was considered as the return for the original expense of improvement. Had continual additions of limestone, gravel, and clay been made to the soil, the expense would have been amply repaid, and the soil would at length have become permanently profitable.

I have also heard it asserted, that ten pounds per acre laid out upon the best land in Ireland would yield more profit, than if laid out upon an acre of Bog, or other unreclaimed land; this is one of those vague assertions, which are made without any foundation in experience, but merely to supply the want of something better to say. If this were the case, why should bills for the inclosure of commons pass every year in England?—Why should not the boundless capital of British merchants raise the value of land near the metropolis tenfold every year?—It is true, that every acre of ground near London might be converted into a dunghil; but though a dunghil and a hot-bed are the most productive soils for asparagus and cabbage, could the butter of Ireland, or the cheese of England, be produced by cows fed on dunghils?—The garden culture of Battersea and Brompton is carried to the highest perfection, and it is profitable; but not to such a degree, as to invite that species of competition, which always follows high profits.

The scheme of improving the Bogs of Ireland is by no means new. The Dutch, in the time of King William, offered, upon condition of being governed by their own laws, to form a colony in the Queen's County, and to make meadow of the whole Bog of Allan.—(*v.* Philosophical Survey of the South of Ireland, page 126.)—In fact, the experience of ages teaches, what Bacon has told us, that great gains are to be made by parsimony, or by new inventions.

It is not surprising, that the project of improving Irish Bogs should have occurred at different periods, both to individuals who had only their own profit in view, and to the patriots who were zealous for the prosperity of their country.

Undoubtedly it is an object of the highest importance to the State, and particularly to this portion of the empire, because the modes of life are such in Ireland, as would immediately be suited to the cultivation of the kind of soil, which may be obtained by the first stage of improvement in Bogs. Much of the cultivation among the great mass of the people in Ireland is carried on by the labour of men without the plough; the tender soil of newly reclaimed Bog might not for some time bear the tread of cattle, though it might be manufactured by the spade and shovel; and a time will come, when many thousand hardy Irishmen must return from the army and navy to their own country, where to find employment of any sort will not be very easy, and where the wages of ordinary labour are much lower than a soldier's pay; but in the cultivation of land, fit for the immediate production of the common food of the country, they would find a resource against idleness; as they would, if they were assisted by the gift of some timber from the proprietors of the Bogs, soon build themselves habitations, and would thus have a reasonable hope of being able to support families. Independently of this more remote speculation, we see, that in this part of the country the population increases rapidly, and that the fathers of families, when their sons grow up, are obliged to divide their small farms among their children; a system, as Lord Selkirk has shewn, in his excellent "Essay on Emigration," to be the most destructive of happiness, that can be followed in an agricultural country, where there are no means of adding to the quantity of profitable land.

But even if the means proposed for bringing our Bogs to this state of cultivation should not ultimately succeed, yet our labours have not been in vain: to have surveyed these Bogs; to have acquired the most exact knowledge of the levels of their surface, of the depth and consistency of the soil, which lies beneath them, of the direction and fall of the streams and rivers, which intersect them, and which may carry off their superfluous water to the sea; is undoubtedly advantageous to the country, as it enables us to determine, either that to reclaim our Bogs is impracticable, and a project too expensive, or, on the contrary, that there are well-founded hopes of success. Further, to enable the nation to decide whether this project be a vain speculation and a *job*, or a useful feasible plan, let a hundred acres of Bog be procured, by rent or purchase, in every district, which has been surveyed under the orders of your Board; and let such plans as your Board approves be fairly put in practice, under the inspection of Engineers of your appointment; let the expense be limited within a certain sum per acre, and the result will be before the public. If it be unfavourable to the scheme, it will at all events be allowed, that the surveys made under the orders of the Board are valuable materials for a topography of Ireland; that the inquiry will have raised a spirit of research and habits of accuracy, that were unknown on such subjects in this part of the United Empire; and that it has introduced into the country a considerable number of excellent instruments, which will oblige surveyors hereafter to study the construction of mathematical instruments, and to be ashamed of such tools as were formerly in use.

I have seen a surveyor of extensive practice employ a circumferentor, the sights of which were tied on with packthread; and another, with *an iron nut and screw,* to one of the legs of his instrument, within an inch of the needle!

Instead of the miserable scrawls, which were formerly given in as maps of gentlemen's estates, none but distinct and exact surveys will now be tolerated. If this improvement in the professional habits of numbers of men of understanding, which had been previously cultivated by a certain degree of science, be not in itself a great advantage, there is no faith in the experience of the past, nor any trust in the hope, that as men become enlightened they become better citizens. But if these plans and these experiments succeed, the benefit will be instantly perceived and permanently felt; no person possessed of Bog can fail to wish for its improvement; and all that can then be wanting would be capital. This might perhaps be safely advanced upon landed security, to landed proprietors, and at such times, and by such degrees, as would leave but little probability of loss, or temptation to fraud. But without the assistance of Government, if we succeed even upon a small scale, British capital will soon be found ready, to carry on such profitable work to a large extent. Those who are tremblingly alive to every interference with their property, even with those parts of it which yield them nothing at present, and give no hope of ever yielding any thing, would embrace eagerly a proposal, that should leave this darling property untouched by any hand but their own, and which should hold out to them a certain prospect of gain.

The improvement of Irish Bogs has been a favourite object with every true patriot, who has written on the affairs of Ireland. There is, in particular, an excellent Essay, written near 150 years ago, upon this subject, in the "Philosophical Transactions," (Lowthorp's Abridgment, vol. ii, page 732), in which there is the following passage:

"*An Act of Parliament should be made, that they who did not, at such a time, make some progress in draining their Bogs, should part with them to others that would.*"

This is strict justice in every nation upon earth. No man, from prescription or from any grant, has a right to monopolize and withhold the soil from cultivation. Could the proprietors of the coal mines and salt pits in England be permitted, to combine and shut them all up? Or could the owners of all the Bogs in Ireland lawfully combine, to prevent them from being cut for turbary?

Fortunately for society, such extreme cases do not often occur. Motives of interest in time overcome the caprice of pride; when mankind clearly perceive private advantage and general utility in any national undertaking, the narrow prejudices of a few are soon despised and overborne; and those who had been most clamorous in their resistance, become the most vehement partizans of the very project, which they had formerly opposed.

RICHARD LOVELL EDGEWORTH.

28th October, 1810.

In part of his report I find the following answer to those who fear that a scarcity of fuel may result from reclaiming the bogs of Ireland.

"An apprehension is generally entertained, that in the event of the improvement of the Bogs, the country would be left without a sufficient supply of fuel. It seems not to be generally understood, that, if the Bogs of Ireland were reclaimed, we should not merely derive the advantage of cultivating their surface, but at the same time the power of applying them, wherever necessary, for fuel, would be augmented some hundred, or rather some thousand fold. Fuel can at present be obtained only from the edges of these Bogs; the excessive wetness of the interior rendering it, in its present state, wholly unavailable for this purpose; but if once drained, fuel might be obtained from every part of them. And it is a great mistake to suppose, that the drainage of a Bog would impair its quality as fuel; on the contrary, it would operate as the greatest possible improvement of it."

My father left means of making all that he had done in examining the Bogs in his district permanently useful to whoever may afterwards carry on their improvement. For this purpose, he availed himself of an ingenious suggestion of his son William's. He had the number of every station, which was bored or levelled, actually cut out in large figures or letters, of about six feet long, on the surface of the bogs, and a heap of turf was placed at each letter, which guided the eye upon the bog, and by following these stations the operations performed at each might be verified and referred to at any time. Marks, referring to their numbers, are laid down on the maps in their proper places; so that after a bog has been surveyed, all the intended improvements, places for drains, &c. may be explained in writing, for workmen or their overseers.

Statement of the comparative profit of land of a good quality employed in farming or planting.

A GROVE, planted on a good strong soil, in the demesne of Edgeworth's-Town, which was planted sixty-three years ago with beech and ash, was measured, numbered, and valued in April, 1809.

The ground, including the measurement of the *backs* (or banks) of all the ditches, on all of which trees grew, amounted to seventy-eight perches. This may, for the sake of easy calculation, be taken as half an acre. Supposing this half acre to be worth five shillings per annum in the year 1745, if it had been then sold at twenty years purchase, and the amount, viz. 5*l*., laid out at compound interest, it would in sixty years amount to 160*l*.

Supposing that the land had been let from seven years to seven years, at five shillings for the first period, and increasing at the rate of one third additional at the end of each seven years, that is to say, at

£	s.	d.	
£0	6	8	for the 2d period,
0	9	0	for the 3d,
0	12	0	for the 4th,
0	16	0	for the 5th,
1	1	3	for the 6th,
1	18	0	for the 7th,
2	10	0	for the 8th, (equal to about 5*l*. per acre),

which is as high as it could well be let for, the amount of the whole would be 63*l*. 5*s*. 7*d*. without reckoning interest; but if the amount of the rent, at the end of each term or period of seven years, were laid by at 6 per cent simple interest, the interest would amount to 61*l*. 6*s*. 6½*d*, making in the whole 124*l*. 12*s*. 1½*d*.

Supposing that the land had been let at three successive periods, on leases of 21 years each, paying for the first 21 years, at 5 shillings for the half acre,

£1 0 0 for the 2d period of 21 years,					
1 10 0 for the 3d, (which would have commenced in 1788),					
The principal sum received would be - -	£63	0	0		
The interest - - - - - - - - - - -	65	0	0		
	128	0	0		

But here it must be observed, that no interest is charged for any thing received during the last 21 years, which would, by a rough statement, amount to 9*l.* 10*s.*

Now the trees which were, this 4th of April, 1809, numbered on this half acre, were 134, (3-4ths of which were beech), and the value estimated at 250*l.*; from this should be deducted the original cost of planting, and the compound interest thereon, amounting to about 50*l.*

The remainder is nearly double what the rent would have amounted to, with the interest carefully laid up, according to either of the foregoing suppositions. It is one quarter more than the amount of the fee simple, improved at compound interest.

Having formerly believed, that it was not profitable to plant good land, though it might be advantageous to plant in an indifferent soil, I thus detail the result of the calculations, which have convinced me of my error—an error, into which I hope none of my posterity, who become possessed of land, will ever be betrayed either by insufficient theory or by neglect.

It should be carefully remembered, that the foregoing calculations admit, that the amount of the fee-simple might be improved at compound interest; or that the amount of rents at the end of periods of seven years, or of twenty-one years, might be laid out at simple interest, and left to accumulate for so long a period as to the end of sixty-three years; not one of which suppositions is in any reasonable degree probable; whereas the profit of planting is a treasure buried in the earth, which increases every year, and which the possessor is not tempted to dig up, till it has accumulated to a large sum.

Extract from Mr. W. B.'s Letter to Mr. E. respecting a Telegraph in India, referred to Chapter XVIII, vol. ii. p. 373.

I recollect, that my brother stated to you the success of our experiment, the substance of our memorial and proposals, and how the last were dealt with by the government.

I have got as many telegraphs ready, as are sufficient for the line at present in question, viz. between Bombay and Poona. I have a corps of a hundred telegraphers, which I raised myself, perfectly taught, and have my vocabularies finished after a very copious manner. I am waiting to proceed for a letter of permission from the Mahratta Chief, through whose territory part of the line is to pass, to erect telegraphs in those places, which are best adapted for stations; the line that I am now about to establish will be no more than a temporary one, and sanctioned by government, merely through my recommending it; and the reason for doing so, you shall hear: what was first proposed to me after the experiment was, to establish a line between this place and Madras at once; but at the next sitting of council, they came to a determination, to make it first known to the government of Madras, that both presidencies might join about the expense. At this time I got a number of telegraphs made, and finding Poona to be in the direction, and a place in which a communication with Bombay would be of consequence, I recommended to them to have the line erected, in the interim of their hearing from Madras, and to place as many men on it, as to afford a tindal or corporal at each station in the long line, on its being finished; the men are to lie in tents at present, which are as convenient at this season as houses, the monsoon being entirely over; but the real line will have a wooden house, and a wooden telegraph at each station; each house is to be of such a size, as to hold twelve sepoys and four telegraphers, and the telegraph is to be of such a fixture, as to be capable of withstanding the severity of the monsoon seasons.

I suppose from the direct distance to Madras, that about

one hundred stations will be required; and I have entered into contract with government to finish it, agreeably to the above plan, for 700 rupees, or 87*l.* sterling per station, cost of telescopes included. My salary, as superintendant, is 600 rupees per month, which is equal to 900*l.* sterling per year; and my brother has a like salary, for managing the communication at Bombay: we are entitled to this from the 28th of August, the day on which I received the first official letter to proceed with it. This may seem to be a business of advantage to me; yet I assure you, that my motives for undertaking it were far from being pecuniary ones; for my brother, in expectation of my assistance as a partner, enlarged his concern to such an extent as to be capable of clearing 6000*l.* sterling per year, exclusive of all expenses; one third of this sum to be mine, and two thirds his. My aims were at celebrity, and to gain a creditable repute for myself in the country, to which (if I may speak my sentiments) any person of spirit ought to sacrifice what is termed *gain.* I am happy to say, I already perceive strong proofs of my having succeeded in my aims; for the Honorable Government, and several of our great folk, from my proposals, have a good opinion of me. The only opposition I have met with, on the whole, has been from some fastidious engineers, an account of which would be needless to give you. I have the pleasure, however, to tell you, that by the great interest my brother has, and the merits of the telegraph *itself,* I have got rid of them, although not without some trouble. * * * * * * * * * * * * * * The inventor of this telegraph being now universally known here, from my memorial, &c., it adds much to that celebrity, which *it* also ought in justice to have procured you in Europe; and you may believe me, my worthy Sir, that I should feel as much happiness in having contributed in any way to that end, as in any credit it might gain for myself.

I shall conclude my accounts of the telegraph, by informing you by what steps I have brought it to perfection here. I was obliged to construct the first myself. It is extremely troublesome, to make the native mechanics understand a person's meaning; but, like the Chinese, they are very clever at patterning—wonderfully so, when one considers the means and methods they have for doing business. As for the telegraphers, I have found them tolerable docile, so far as relates to making signals, but exceedingly hard to be taught the managing a *middle station;* however, I have with much ado made about thirty quite tractable, even in this. In order to avoid crowding their minds (which, I must say, are very confined) with a multiplicity of signals, I have laid aside distinguishing the words of the Appendix by vibratory motions, and have chosen rather to submit even to the inconvenience of an increase in the numeral index, than to begin the Appendix with unity; as I consider the time spent in vibrating at a figure, would not be wanted to express one. Pardon me for this remark, as I do not propose it by way of amendment, but only on the above account, viz. the stupidity of the men. I find, that, without encreasing the index to more than five places of figures, notwithstanding the destruction there are still retained 10908 numbers. I should suspect then that seven thousand words, selected with judgment, ought to be copious enough for the vocabulary; and the remaining 3908 are nearly enough for the Appendix, which is composed of the hours of the day, the names of the most conspicuous personages, the names of places of note, a list of his Majesty's ships' stations in India, the name of the Honorable Company's marine ships, ditto of the ships trading between England and India, called Indiamen, and finally, of about five hundred sentences, which I have since taken singular pains to compose; the whole arranged in alphabetical order, and in my opinion sufficiently powerful, to answer every purpose, that can in reason be expected from such an institution.

On my way from England, I applied myself with the strictest attention to navigation, (of which I had previously a tolerable theoretical knowledge,) and to the lunar observations, until I became master of both; and since I have been here, my brother has taught me the use of that beautiful instrument, the plane-table.

Seeing that I have got such a field open to exercise my talents, my intention is to avail myself of it so far, as to take a survey of the line between this place and Madras, ascertain the latitude and longitude of the places of most note which intervene, and keep a journal of remarks on the population, &c. &c.: these perhaps may prove useful; if in no other respect, they will be a pleasure to myself.

I dread I have extended this beyond the bounds of patience, but at commencing I determined to write in detail; as the subject on which it is chiefly written is that, which I know you to have that regard for, which its excellence justly deserves.

W***** B****.

IN the preceding Memoirs, some lines of poetry of my father's have appeared. He had peculiar facility and quickness in turning any ideas, especially any epigrammatic thought into verse; and though he undervalued this talent, and certainly never pretended to be a poet, I will venture to give a few specimens in different styles of his *vers de société;* such they may in truth be called, as most of the lines were actually spoken or written in company.

EPIGRAM

On receiving Dugald Stewart's Works in exchange for "Practical Education."

BY Pallas taught, Tydides learnt to pass,
In fair exchange for gold, his shield of brass;
But Edgeworth, wiser than the chief of old,
Without Minerva palms his lead for gold.

1796

EPIGRAM

On some recent Scotch Marriages and Divorces.

To ready Scotland boys and girls are carried,
Before their time, impatient to be married.

Soon wiser grown the selfsame road they run,
With equal haste, to get the knot undone;
Th' indulgent Scot, where English law too nice is,
Sanctions our follies first—and then our vices.

Feb. 1811.

EPIGRAM.

On a picture given by Henry IV *to the Chevalier D'Aubigny, who had done him essential services.*

C'EST un Roi d'étrange nature,
Je ne sais qui Diable l'a fait,
Car il récompense en peinture
Ceux qui l'ont servi en effet.

Translation.

BEHOLD how services by Kings are paid!
They take in substance, and they give in shade.

October, 1796.

ORDERS TO MY PORTER.

(Imitated from the French.)

THOU faithful guardian of these happy walls,
Whose honest zeal protects thy master's gate:
If any stranger at this mansion calls,
I'll tell thee who shall enter, who shall wait.

If Fortune, blindfold dame, should chance to knock,
Or proud Ambition court me to her arms,
" Shut, shut the door, good John," and turn the lock,
And hide your master from their Syren charms.

For in their dismal train, as black as night,
Come, hideous Care, and sullen Melancholy;
And Song, and Joy, and Laughter, take their flight,
Nor leave one precious moment to dear Folly.

If at my door a beauteous boy be seen,
("His little feet have oft my threshold trod,")
You'll know the offspring of the Cyprian queen,
His air, without his bow, betrays the god.

His magic smiles admission always win,
Though oft deceived, I love the dear deluder;
Morn, noon, or night be sure to let him in,
For welcome Love is never an intruder.

Should sober Wisdom hither deign to roam,
Nor let her in, nor drive her quite away—
Tell her at present I am not at home,
But hope she'll visit me another day.

On receiving a Pencil-case from Mrs. E. Edgeworth, with a Black Lead Pencil at one end, at the other a gold pen.

If in some heedless hour my careless strain
Should chance to give my loved Eliza pain,
May the rude lines the fading pencil trace!
May the rude lines her gentle hand efface!
But when her worth, or when my love is told—
Oh! may the sterling line be graved with gold.

The two following playful enigmas were written in the intervals of violent pain, in his last illness; one of them is that which is referred to in his letter to Lady Romilly, p. 446.

"The sages say that every thing
Must from mere form and matter spring,
Be the materials what they will,
From *form* they take their value still—
From *form* my influence I receive,
And what is given to me, I give;—
My power is such as to combine
Things most unlike, and make them join;
The mountain goat, and lowland sheep,
Mix with the monsters of the deep;—

The silkworm and the ox I range,
Together in a union strange!
The world of plants its tribute sends,
And Art its glad assistance lends.—
Who would not think, if this be true,
The foremost honours were my due?—
Yet from my rank I'm always cast—
The *first* in use, in name the *last.*— *May,* 1817.

ENIGMA.

I'm something, I'm nothing, I'm both, or I'm either;
Tho' some folk there are, who will tell you I'm neither;
Though bound to no party, I always support
The designs of the schemer in country or court:
When the job is completed, they make it their pride,
To pull me to pieces, and throw me aside;
Though things move around me for ever and ever,
To fly further off, is their constant endeavor;
But such is my power, let them strive as they will,
To me they are drawn irresistibly still.

LINES

Addressed to my dear children, in my 73d Year; when in declining health, and my sight nearly lost.

With boys and girls, a baker's dozen,
With many a friend, and many a cousin:
The happy father sees them all
Attentive to his slightest call;
Their time, their talents, and their skill,
Are guided by his sovereign will,
And e'en their wishes take their measure
From what they think the patriarch's pleasure:
" How does he rule them?—by what arts?"——
He knows the way to touch their hearts.

List of essays of Mr. Edgeworth's, which have appeared in different periodical publications.

Philosophical Transactions.

Essay on the Resistance of the Air. . . . vol. lxxiii.—1783.

Account of a Meteor. . . . . vol. lxxiv.—1784.

Transactions of the Royal Irish Academy.

Essay on Springs and Wheel-Carriages. . . vol. ii.—1788.

Essay on the Telegraph. . . . . vol. vi.—1795.

Monthly Magazine.

On the engraving of Bank of England Notes, vol. xii.—1801.

Nicholson's Journal.

Essay on Rail-Roads. . . . . vol. i.—1801.

Description of an Odometer for a Carriage vol. xv.—1806.

Remarks on Mr. Ryan's Boring-Machine. vol. xv.—1806.

On the Construction of Theatres. . . . vol. xxiii.—1809.

On Telegraphic Communication. . . . vol. xxvi.—1810.

On roofing Longford Jail with flag-stones vol. xxix.—1811.

Description of a new Spire. . . . . . vol. xxx.—1811.

On Portland Stone, as a covering for the Spire, vol. xxxi.—1812.

On Aerial Navigation. . . . . vol. xlvii.—1816.

On Wheel-Carriages. . . . . vol. xlviii.—1816.

Letter from Z. . . . . vol. xlix.—1817.

The following are fac similes from the Sketches alluded to Chap. xiv.

No. I.

This *Traiteur* is a Portrait taken from the Life at Paris in 1802.

C. E. del.

F. C. Lewis. scu[lp]

Le garçon traiteur.

Published March 30, 1820, by Rowland Hunter, St. Pauls Church Yard

C. E. del. G. Lewis, sculps.

The Knave.

Published March 30, 1820, by Rowland Hunter, St Pauls Church Yard

No. II.

The Knave and Slave were also drawn from the life in Ireland. The names are given merely as suiting the expression of the countenance and figures. There are large classes in Ireland, of which these portraits are characteristic.

The Knave was not absolutely dishonest; he was only one, who, with slight variation, adopted the Negro's maxim of " God gives black man all that white men forget:" reading it, God gives poor men all that rich men forget. He was a man of much resource, humour, and wit, not restrained by strict regard to truth. Indeed, without any conscience upon this point. This was the person, who described the array and vestments of the fairies. He averred, that he had *visibly seen* the good people, as he called the fairies, dancing on the grass in mid-day. He called us to look at them, declaring, that he saw them plainly, even while we looked to the spot to which he pointed. (Mentioned in the notes to Castle Rackrent.)

No. III.

The Slave.—This appellation does not imply, that the man was enslaved. A slave formerly in England and still in many parts of Ireland simply means a man, who earns his subsistence as a day-labourer. Some years ago, if a labourer or working man was asked what he was, he would answer, " I am a poor slave"—To slave, being synonimous for to work hard.

The individual here represented did not *slave himself* much, for he was, as might be supposed from his portrait, lounging in his gait, slow in all his motions, and lazy in all his habits. But as to the rest, he was an honest, affectionate creature—in his youth a sportsman, and an excellent shot. The happiest hours of his life, and those in which he gloried in his later days, had been spent in going out shooting with his master when a boy—to whom, and to his family, he ever continued faithfully and strongly attached. He survived him but a few months ; he was carried off by the fever last year (1818).

C. E. del. G. Lewis.

The Slave.

Published March 30, 1820, by Rowland Hunter, St Pauls Church Yard

C. E. del.

The Thief

Pubd. ch. 30. 1820, by Rowland Hunter.

F. C. Lewis, sculp.

The Witness

No IV.

THE THIEF AND HIS WITNESS.

The Boy, who holds his hat before his mouth, had good reasons for wishing to hide the expression of his countenance. He had robbed his grandmother of sixty guineas, which she had kept concealed in an old flower-pot.—The witness was an accomplice in the theft.—He was, as his leather apron bespeaks him, a shoemaker, or as he would have said of himself, "*partly* a shoemaker."

One slight sketch of these two boys was taken in the attitude in which they stood—the thief foremost, his witness behind him, when they were brought before my father.

No. V.

The *Irish Dance* is an original composition. None of the figures are, as far as I know, portraits. This sketch represents a scene common in Ireland—one of their dances called patterns (*patrons*), from the *patron* saints—each Catholic having a saint for his patron.

FINIS.

C.E. del.
Published M

Irish Dance.

W. Cox, sculp.

by Rowland Hunter, St Pauls Church Yd

Directions to the Binder for placing the Plates.

VOL. I.

VOL. II.

J. M'Creery, Printer,
Took's Court, London.

Printed in the United States
By Bookmasters